16035
£7.99
TI S435

Number Freaking

```
1111100001110111000111111000111000001011111011100001011101000001000001111
1111011111101011011011111011011011111011111011011111011101101110110111111111
1110110001011010110111110110110100001101111101011000100010000011101100001111
1111011011010101101101111101101101101111011101101011010110111011011111111011111
1111100001110111000111111000111000001000001011100001011011100110110011000001111
1111111111111111111111111111111111111111111111111111111111111111111111111111
0101110111111011100011100011111101101011010110101100001100000100001100000
0100110111111010110110110110111110011010110110100100101110101111101110101111
0101010101110110101101011101110101010101011010101010101000001100001100001100000
0101100111101010110110110110111110110010110101101011011010111110101011111110
0101110111111011000111000111111011101000001011101000011000001011101001000000
1111111111111111111111111111111111111111111111111111111111111111111111111111
111111111111111111111111111111111111111111111011010101000011110000010101111
11111111111111111111111111111111111111111111011101010101101101111101011111
1111111111111111111111111111111111111111111111010101001011010110010010101111
1111111111111111111111111111111111111111111111010110101010101110101101010111
1111111111111111111111111111111111111111111111011101011101110000010100000
```

Number Freaking

The Surreal Sums Behind Everyday Life

Gary Rimmer

Icon Books

This edition published in the UK in 2006
by Icon Books Ltd., The Old Dairy,
Brook Road, Thriplow, Cambridge SG8 7RG
email: info@iconbooks.co.uk
www.iconbooks.co.uk

Originally published in the UK and Australia in 2005
under the title *How to Make a Camel Smoothie*

Sold in the UK, Europe, South Africa
and Asia by Faber and Faber Ltd.,
3 Queen Square, London WC1N 3AU
or their agents

Distributed in the UK, Europe, South Africa
and Asia by TBS Ltd., Frating Distribution Centre,
Colchester Road, Frating Green, Colchester CO7 7DW

This edition published in Australia in 2006
by Allen & Unwin Pty. Ltd.,
PO Box 8500, 83 Alexander Street,
Crows Nest, NSW 2065

Distributed in Canada by
Penguin Books Canada,
90 Eglinton Avenue East, Suite 700,
Toronto, Ontario M4P 2YE

ISBN 10: 1-84046-751-7
ISBN 13: 978-1840467-51-2

Text copyright © 2005 Gary Rimmer

The author has asserted his moral rights.

Typesetting by Hands Fotoset

Printed and bound in the UK by Bookmarque Ltd

Contents

About the Author

Gary Rimmer is a journalist, documentary television producer and award-winning screenwriter. As well as writing for the national press he has contributed to magazines ranging from *Time Out* to the *British Medical Journal*. He is the author of *Lonely Hearts* and *Thirtysomehow* and co-author of *The A–Z of Street Cred*. He has been there and done that. He hopes to live to the age of 3,518,791,200 seconds.

You Do the Math

Nothing lasts: not this book, not you, not even the universe. Best not to think about it too hard on a Monday morning, but eventually everything turns to dust. And when all is gone? All that will be left will be what some old dead Greek called the music of the spheres ...

But just in case you don't live for a few billion more years, you can hear this merry tune now at www.astro.virginia.edu/~dmw8f/index.php, where astronomer Mark Whittle has compressed the cosmic background radiation from the first million years after the Big Bang into an audible sound file. Feel free to sing along.

Sticking with the philosophical for a moment, you could say that what Professor Whittle has created is the sound of numbers talking. Numbers are the invisible sinews shaping our reality, the meta-text of all we survey. In a postmodern world where no truth can be considered absolute any more, sometimes their elegance and precision are all we have.

However, sometimes numbers can be too clever for their own good, the rules that govern their manipulation too immutable and eternal. Think about your local shop. Let's say a tin of beans costs £1 or $1 or whatever the currency is in your preferred country. Apart from being overpriced, how much will three tins cost? £3? $3? No, because the shop has a three-for-two offer on baked beans today. Sometimes numbers aren't as accurate as they'd like you to think.

It's with this thought in mind that you should read *Number Freaking*. This isn't a book of statistics and it isn't a book of science. It has nothing whatsoever to do with actuaries or accountants. *Number Freaking* is a book of surreal sums and absurd arithmetic: it's doodling with numbers, doing sums in your head just for fun, playing dice with the universe. Number Freaking is the art of putting numbers where none existed before to take an off-the-wall peek behind the curtains at how numbers rule our lives. Number Freaking reveals the low drama of life, the unexpected realities and unforeseen truths that emerge only when numbers are tested to destruction.

Anyone can do it; everyone can enjoy it. Think of this book as a sampler. With calculator in hand, or simply a fiendish way with mental arithmetic, you can work out the answers, or skip to the end of each page for every solution. Sometimes a wild guess is all you'll need when Number Freaking, but a maths nut is the last thing you need be to come along for the ride.

Number Freaking reveals the world, through the lens of numbers, in a different light. It isn't about finding right answers because there may not be any. It isn't even about accuracy: it's about averages and assumptions, ballparks and coincidence, extrapolations and guesstimates. This book was inspired by a single statistic – how many Chinese smokers there are – and an unhealthy desire to work out how many cigarettes they smoke each. But what you Number Freak is up to you. Of course, you can make up your own. You probably do make up your own.

All the Number Freaks in this book can be answered with basic arithmetic and the 'facts' included. Every number in the book is real, and taken from everyday life. And every effort was made to ensure these numbers were as correct, accurate and up-to-date as was possible at the time the book was written. They're derived from hundreds of sources. But here's a friendly warning: they shouldn't be regarded as an authoritative source. Well not in court anyway. But over drinks with your friends? That's up to you ...

And lastly, always remember the Number Freaking motto: 'God may not play dice with the universe, but Number Freakers do.'

So Let's Go Number Freaking ...

Basic addition, subtraction, multiplication and division are all the skills you'll need to go Number Freaking. Most of the sums in this book were calculated on a battered eight-digit Texas Instruments calculator. Consequently, if you work long-hand, or if you've got a better calculator, or you decide to use the excellent on-line nine-digit calculator on Google instead, some of your answers may work out to be a little different. So if you do find a number that's wrong, thanks, but don't bother getting in touch ...

As for how many decimal places answers are expressed to, and indeed whether they've been rounded up or down, this was dictated entirely by the question. Whether it seemed more interesting to go to two instead of four, one instead of three, or three instead of none, was decided entirely on merit.

Finally, a note about conversion factors and conventions. Many conversion factors are absolute, or fixed by international convention. So an inch is 2.54 centimetres, a kilo is 2.205 pounds, and a mile is 1.6093 kilometres. The values used for big numbers – a billion is taken to be a thousand million and a trillion a million million – are those accepted by scientists worldwide. (Other conversion factors and conventions you can probably find in any half-decent diary.) Where the UK and the US find themselves divided by a common measure, the UK measure has been used. So a pint is 0.5683 litres. (A gallon is, of course, 8 pints.) However, some numbers change. For example, in the time it took to write this book, the global population increased by almost 80 million. Where numbers are evolving, they are fixed at their value when this book was completed. So the population of the world is fixed at 6.388 billion, and the conversion factor to change UK pounds into US dollars at £1 to $1.7966. Likewise, the wealth of the mega-rich (and that of mere mortals), where quoted herein, is that which was being reported at the time the book was written.

One important pair of conventions adopted throughout this book is that a month is 30.25 days (on average) and, to allow for leap years, a year is 365.25 days.

Anything else you need to know? Three things. First, you'll be told when assumptions are made. Second, thanks are due to Bronwen, Sam and Ruby for several of the reasons authors usually put in acknowledgements. Third, you may well have a calculator on your mobile phone ...

Why Did Elvis Really Die?

By the age of 42 when he died of heart disease, Elvis Presley was eating 65,000 calories daily. In the UK the recommended daily maximum for a man is 2,550 calories.

i. So by the time he kicked off his Blue Suede Shoes, Elvis was chomping his way through the maximum daily calorie allowance of how many British men?

a) 17.76
b) 25.49
c) 44.14

According to McDonald's, an American Big Mac sandwich contains 600 calories. 4.25 Big Mac sandwiches therefore contain 2,550 calories.

ii. Elvis was eating the equivalent of how many Big Macs daily when he died?

a) 54.00
b) 142.23
c) 108.33

To burn off 3,235 calories takes an 8 mile cycle ride at 17 mph.

Answers: i) b ii) c

A regular McDonald's hamburger sandwich contains 280 calories.

iii. How far would Elvis have needed to cycle at 17 mph to burn off his 65,000 calorie daily intake?

a) 165.25 miles (266 kilometres)
b) 160.74 miles (259 kilometres)
c) 155.50 miles (250 kilometres)

- *Distance London to Swansea: 160 miles*
- *Distance Miami to Key West: 160 miles*

What killed Elvis was a chronic lack of cycling.

A Punch from Ali: Bee Sting or Car Crash?

In an interview with Michael Parkinson, Muhammad Ali claimed he could square up to Parkinson, hit him, and get his hand back to where it had started in the time it would take Parkinson to blink.

Assume this is true.

Ali is recorded as 6 feet 3 inches or 75 inches (1.905 metres) tall. According to the 'Golden Ratio', this *should* make the Louisville Lip's reach 75 inches. However, Ali's reach reportedly measures 83 inches.

According to the science of *anthropometrics*, these body dimensions give Ali a shoulder to knuckle length of 99.53 centimetres. We'll call it a metre. (So a punching length of 1 metre one way, 2 metres there and back.)

The blink reflex lasts 0.25 seconds.

Answer: iii) b
Your mouth produces 1 litre of saliva a day.

i. At what speed was Ali claiming his fist could move?

a) 28.80 kph (17.90 mph)
b) 92.86 kph (57.71 mph)
c) 118 kph (73.34 mph)

In fact this is an average speed. In reality, to be faster than a blink Ali's fist would have to accelerate to at least this speed in an eighth of a second, stop dead, then do the same coming back. If somehow his fist could accelerate continuously, by the time he'd pulled it entirely back, it would be travelling at 86.4 kph (53.7 mph).

* *About 35 per cent of 50 mph car crashes result in a fatality.*

Being punched by Ali was like being hit by a car. And you wouldn't have seen it coming.

Celebrity House Values

These are the prices paid for their London homes by a number of celebrities of various standing:

Hugh Grant: £2.10 million ($3.77 million)
Madonna: £5.77 million ($10.37 million)
Patsy Kensit: £1.38 million ($2.48 million)
Sven-Goran Eriksson: £2.55 million ($4.58 million)
Tracey Emin: £0.91 million ($1.63 million)

Actually, in terms of celebrity, this lot are at least 'B' List. If we take what they paid as indicative, the average sets some kind of benchmark ...

Answer: i) a
A human brain weighs 1,500 grams.

i. So if you claw your way up the greasy pole of fame to the heights of an 'A/B' List celeb, what should you pay for your London *pied à terre*?

a) £1.11 million ($1.99 million)
b) £3.29 million ($5.91 million)
c) £2.54 million ($4.57 million)

So Sven has got it just about right.

The average price of a house in the UK is currently £177,474 ($318,850).

So to be a celebrity you must be prepared to spend about fourteen times more than an average person on a house. Do you think you could do that?

Humungous

These are the names for multiples of ten agreed by the world's scientists.

Number of zeros	Name	Number of zeros	Name
1	ten	33	decillion
2	hundred	36	undecillion
3	thousand	39	duodecillion
6	million	42	tredecillion
9	billion	45	quattuordecillion
12	trillion	48	quindecillion
15	quadrillion	51	sexdecillion
18	quintillion	54	septendecillion
21	sextillion	57	octodecillion
24	septillion	60	novemdecillion
27	octillion	63	vigintillion
30	nonillion	100	googol

Answer: i) c

In the US Santa Claus receives 627,000 Christmas letters.

Multiples of these numbers are expressed in the same compounded form as smaller numbers, for example one hundred thousand or ten octillion.

i. Using 'one hundred' as an adjective, how many zeros would there be in the longest number – in terms of the number of letters – that can be named from the list above?

a) 23
b) 62
c) 47

ii. Using 'one hundred' as an adjective, how many zeros are there in the largest number that can be written, where the number of numbers is equal to the number of letters when it is written down?

a) 20
b) 24
c) 27

iii. What is the largest number that can be written, where the number of numbers is equal to the number of letters when it's written down (without using words like 'and')?

a) 923 sexdecillion
b) 878 nonillion
c) 345 trillion

If in doubt you can always say zillion.

Answers: i) c ii) a iii) b

A googolplex is 10 raised to the power of googol.

Blood Suckers

The biggest insects can be almost prehistoric in their scale: apart from stick insects as long as a man's calf, there are beetle species as big as billiard balls – and three times the weight of a mouse. And not all of them would you want at your birthday party ...

The most voracious blood sucking insect on earth is reported to be the female bont-legged tick *Hyalomma asiaticium*. Native to Central Europe, it can slurp up to 8.856 millilitres of its victim's blood (usually grazing sheep or cattle) in one sitting. An average human has about 5.2 litres of blood in their body.

> **i. Sucking simultaneously how many of these ticks would it take to suck a man dry?**
>
> a) 170
> b) 587
> c) 1,346

Now fortunately female bont-legged ticks do not swarm. Let's hope they never do. A locust swarm might contain 28 billion animals. If the female bont-legged tick ever follows suit, humanity is toast. A female bont-legged tick swarm the size of a swarm of locusts could pretty well exsanguinate 80 per cent of the UK population simultaneously.

Sickly Sweet

We all know scientists need huge numbers for astronomy and computing.

Answer: i) b

The smallest insect is now thought to be *Megaphragma caribea* from Guadeloupe at 0.17 mm long.

Another use is the study of bugs. Bacteria grow at an astonishing rate. The number in a colony can double in twenty minutes. So if you become infected by some bug, the number of pathogens in your body might double every twenty minutes (if it wasn't for the sterling efforts of your body's natural defences).

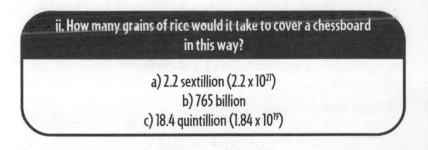

i. How many bacteria might you theoretically have in your body within 24 hours of an infection by one bug?

a) 4.72 sextillion (4.72×10^{21})
b) 8.58 trillion
c) 382 million

Let's try to put this number in perspective. There's a legend concerning the origins of chess in which a wise man suggests that a king reward him with one grain of rice on the first square of a chessboard, two on the second, four on the third, and so on, continuing to double the number of rice grains on each square, until the last square, the 64th square. The king agrees. The wise man ends up with half his kingdom.

ii. How many grains of rice would it take to cover a chessboard in this way?

a) 2.2 sextillion (2.2×10^{21})
b) 765 billion
c) 18.4 quintillion (1.84×10^{19})

The legend talks of grains of rice; in some versions it's grains of corn. We'll modernise it and change those grains to sugar.

Answers: i) a ii) c

The number of bugs would match the entire human population in about ten hours 51 minutes.

There are reputedly 2 million grains in one kilo of granulated sugar. Which, possibly erroneously (but who cares, we're Number Freaking), we'll assume is produced from exactly 1 kilo of raw sugar.

iii. How many metric tonnes (1,000 kilos) of sugar would you have to squeeze onto a chessboard to fulfil the legendary proposition?

a) 45.2 billion metric tonnes
b) 6.7 billion metric tonnes
c) 9.2 billion metric tonnes

- *Global raw sugar production 2003: 145 million metric tonnes*

(And thus, under the terms of the legend, the entire annual global sugar supply would run out on the 59th square of a chessboard.)

If the aforementioned bacteria were grains of rice (or sugar) on a chessboard, then the number of bacteria in your body would match the *total* number of grains trying to squeeze onto our legendary chessboard in about 21 and a half hours.

How to Imagine a Really Big Number

The earth weighs 5.974×10^{21} tonnes

The weight of a grain of sand ranges from 0.3 milligrams for fine sand to 13 milligrams for coarse sand. The very finest grains may weigh as little as 10 micrograms. There are 1,000 milligrams – which is 1 million micrograms – in one gram.

Answer: iii) c

The world's fastest clock ticks a quadrillion times a second.

i. How many grains of 10 micrograms would it take to add up to the earth?

a) 5.974×10^{32}

b) 5.974×10^{28}

c) 5.974×10^{23}

In the language of very big numbers this is 597.4 nonillion.

ii. How many grains of 0.3 milligrams would it take to add up to the earth?

a) 1.9913×10^{11}

b) 1.9913×10^{27}

c) 1.9913×10^{31}

In the language of very big numbers this is 19.9 nonillion.

iii. How many grains of 13 milligrams would it take to add up to the earth?

a) 4.595×10^{29}

b) 4.595×10^{24}

c) 4.595×10^{22}

In the language of very big numbers this is 459.5 octillion.

Clearly all these numbers are beyond human comprehension ... and yet, in the number of grains of sand it would take to 'build' the earth, we can glimpse something of their scale. Imagine yourself flying; imagine how the land below seems to go on for ever. Then imagine a beach somewhere, feel the sand

Answers: i) a ii) c iii) a

A million bacteria can stand on the head of a pin.

falling through your fingers. And then compare the two thoughts. Look, there it is, just a glimpse ...

The biggest number on the previous page – 597.4 nonillion – is roughly the same as 2^{109} (2 raised to the power of 109). A population of bacteria doubling every twenty minutes will reach this magnitude in just over 36 and a quarter hours.

Boom Diddy Boom

An adult human heart pumps 5-6 litres of blood every minute. Call it 5.5. Assume adulthood lasts 63 years (from the age of 15 to 78). The global population is 6.388 billion. Seventy per cent of the earth's population are adults.

i. How many cubic kilometres of blood will all the adult hearts on earth pump between them in 63 years? (There are a trillion litres in one cubic kilometre.)

a) 814,927.37
b) 588,345.22
c) 350,334.66

This is almost equivalent to the combined volumes of the following seas ...

- The Black Sea (508,178 cubic kilometres)
- The Red Sea (214,473 cubic kilometres)
- The North Sea (51,784 cubic kilometres)
- The English Channel (48,546 cubic kilometres)

Answer: i) a

Sixteen million tonnes of rain fall to earth every second.

Filling a Lake? Use a Waterfall

The average flow of water over Niagara Falls is 6,000 cubic metres per second.

Ullswater, in the British Lake District, beside whose shores William Wordsworth saw his famous daffodils, has a capacity of 223 billion cubic metres.

i. How long would Niagara Falls take to fill Ullswater in years and days?

a) 1 year 65 days
b) 1 year 224 days
c) 2 years 16 days

Think of it like a tap filling a bath.

But How??

Apparently 45,000 Americans are injured by toilets every year.

i. Every how often, on average, is an American injured by a toilet?

a) 18 minutes 12 seconds
b) 11 minutes 42 seconds
c) 4 minutes 19 seconds

Answers: i) a i) b

The current at Niagara Falls just before the drop is 20–22 mph.

Data is unavailable on how many victims of malicious toilets are men and how many are women. Since women are more likely to sit, are they more vulnerable to collapse? Men are probably more likely to kneel, so do they have a greater risk of a blow to the back of the head? Or some dental injury? Feel free to speculate.

How Many Eyeballs?

A piece of jargon advertisers use to describe the relative success of their adverts is how many 'eyeballs' it got. In other words, how many people looked at it.

There are 6.388 billion people on earth. Technically they should have two eyeballs each. (Obviously.)

i. What should the global total of eyeballs be?

a) 12.560 billion
b) 12.776 billion
c) 12.924 billion

An eyeball weighs about 7.5 grams. A metric tonne is 1,000 kilograms.

ii. How many metric tonnes should all the eyeballs on earth weigh?

a) 93,229 metric tonnes
b) 102,345 metric tonnes
c) 95,820 metric tonnes

Answers: i) b ii) c

Seven per cent of Americans think Elvis is still alive.

An eyeball is roughly spherical, about the size of a ping-pong ball, and has an average diameter of about 1 inch. It has a volume, therefore, of about 8.58 cubic centimetres. 116.5 eyeballs would therefore occupy a volume of 1 litre.

iii. What should the total volume of eyeballs on earth be?

a) 203.22 million litres
b) 85.91 million litres
c) 109.67 million litres

An Olympic-sized swimming pool has a capacity of the order of 1.5 million litres.

iv. About how many Olympic-sized swimming pools would all the world's eyeballs fill?

a) 141
b) 60
c) 73

Yuck.

I Camera

A recent TV ad for a retail chain of UK opticians claimed we each 'capture' 24 million images with our eyes in a lifetime. Life expectancy in the UK is 78 years.

Answers: iii) c iv) c

Worldwide 180 million people are visually impaired. 40–45 million are blind.

i. If we each saw only 24 million images in 78 years, how long would each image last?

a) 190.8 seconds
b) 38.3 seconds
c) 102.6 seconds

At this rate a driver in urban traffic travelling at 15 mph would update what they saw only about once every 750 yards ...

Of course, in reality a human eye is nothing like a film camera, but it's a useful analogy. A film camera captures images at 24 frames per second (fps). Screen a movie at this rate and movements look seamless and natural. Video cameras do the same at 25 fps. However, a human eye can detect images at a far faster rate. Military radar operators are trained to detect images as transient as one 600th of a second. Normally, though, the maximum refresh rate of the human eye is taken to be 60 frames per second, and the average refresh rate to be about 30 frames per second.

ii. Assume 16.9 waking hours a day, for 78 years, and ignore blinking. At 30 frames a second how many images could you *really* capture through your eyes in a lifetime?

a) 18,776,054,000 (19 billion)
b) 51,999,035,000 (52 billion)
c) 94,223,903,000 (94 billion)

(Which if each is worth a thousand words corresponds to an awful lot of books!)

Answers: i) c   ii) b

At 10 inches the naked eye cannot see objects smaller than 0.1 mm in diameter.

iii. At 30 fps how long would it take to capture 24 million images?

a) 350,000 seconds (5.75 days at 16.9 waking hours per day)
b) 1,800,000 seconds (29.58 days at 16.9 waking hours per day)
c) 800,000 seconds (13.15 days at 16.9 waking hours per day)

So perhaps the TV ad claiming viewers will 'capture' 24 million images in a lifetime should have added: provided they die within a fortnight ...

Scientists are only just beginning to understand human vision. The latest view is that rather than resembling one camera an eye is more like several – about a dozen – separate cameras working simultaneously, each collecting different information about the visual world, which the brain then blends together to create an image. One could perhaps liken it to a printer's colour separations, combining to make a picture. And since each of us has a unique and different brain, we mix these 'inputs' in subtly different ways. It may be physically as well as emotionally true when we say of an artist, 'She sees the world differently to other people.'

Thanks Auntie

In 2003 the BBC spent £2.36 billion ($4.24 billion) on programmes.

The UK population is 59.66 million.

i. So how much did the BBC spend on programmes per head?

a) £106.33 ($191.03)
b) £75.16 ($135.03)
c) £39.56 ($71.07)

Answers: iii) c i) c

Fifty-two per cent of households own at least five remote controls.

Since a colour TV licence costs £121 this suggests that over two-thirds of the BBC's income does not go on programmes. But this is unfair because you only need one licence per household – of which there are 24.83 million in the UK.

ii. So how much did the BBC spend on programmes per household?

a) £95.05 ($170.77)
b) £190.67 ($342.56)
c) £135.33 ($243.13)

Which, at 78.5 per cent of the licence fee, feels like much better value.

See You See Me Tee Vee

So you've died and you're sitting with Saint Peter at the Pearly Gates Security Centre. 'I'm going to show you a film of your life', says he. 'So we can review your case and decide what to do with you ...'

How long might a film of your life last? Well, technically, as long as you were alive. But where's the footage going to come from? Omnipresent or not, even a divine videographer must need to change the tape sometime.

Another alternative is CCTV footage. An average person in the UK now appears in CCTV footage 300 times per day. Notwithstanding moments of fancy editing, a usable 'shot' for television needs to last at least three seconds. Assume every CCTV appearance is three seconds long.

i. How many minutes a week does an average Briton appear on CCTV?

a) 90 minutes (1 hour 30 minutes)
b) 105 minutes (1 hour 45 minutes)
c) 120 minutes (2 hours)

Answers: ii) a i) b
The BBC has a taxi bill of £11.4 million.

ii. How long does an average Briton spend on CCTV in an average
365.25-day year?

a) 5 days 6 hours 8 minutes 7 seconds
b) 4 days 8 hours 54 minutes 50 seconds
c) 3 days 19 hours 18 minutes 45 seconds

iii. So how long in 60 years?

a) 29 weeks 1 day 2 hours 8 minutes
b) 32 weeks 4 days 6 hours 45 minutes
c) 42 weeks 3 days 6 hours 18 minutes

Now we are all minor celebrities on CCTV. Fifteen minutes a day is comparable
to the on-screen presence of a weather forecaster. And poor Saint Peter has
to sit through every second.

Give Us Your Money

Every year, technically, 24.83 million UK households spend £121 each on a
colour TV licence. Companies advertising on TV, meanwhile, spend £4.326
billion, an average of £174.22 per year per household, or 48 pence per day,
trying to sell us stuff on the TV.

There are about 109 million US households. In the USA the total amount spent
on advertising every year is about $248 billion, of which the amount spent
annually on television advertising is about $57 billion.

- *Estimated net worth of the Walton Family (owners of WalMart): $54.2
  billion*

Answers: ii) c   iii) b
British people spend an average of 2 hours 28 minutes a day watching TV.

> **i. How much do advertisers spend on TV advertising per year per US household?**
> (The exchange rate throughout this book is assumed to be £1 = $1.7966.)
>
> a) $972.36 (£541.22)
> b) $745.89 (£415.17)
> c) $522.94 (£291.07)

> **ii. How much is spent per day per US household?**
> (Last reminder, don't forget to assume an average year is 365.25 days.)
>
> a) $1.43 (£0.80)
> b) $2.56 (£1.42)
> c) $4.78 (£2.66)

The American figure per household per day is almost double the British figure. Yet we can assume American TV advertising is at least as effective as British. What can explain the difference? Perhaps Americans must sit through twice as many TV ads as Brits.

Or maybe Americans are just less gullible.

Princess of Pop?

When the Princess of Hearts bared her soul to Martin Bashir, 22.35 million people in the UK watched it on TV.

By comparison, when pop singer Michael Jackson bared *his* soul to Martin Bashir, 14 million people in the UK tuned in.

- *Audience,* Only Fools and Horses Christmas Special, *1996: 24.35 million*

Answers: i) c ii) a

In 2003 there were 1,340 commercial TV stations in the USA.

> **i. So for every five people who peeped at Jackson's dirty laundry, how many took a peek at Diana's?**
>
> a) 8
> b) 18
> c) 180

The controversy surrounding both their lives has been written of extensively elsewhere, so what fresh insights can Number Freaking bring to the debate?

Jackson has sold 210 million records worldwide – more than any other living recording artist. Princess Diana, meanwhile, never had a career as a pop singer.

But imagine she had. And that her sales, relative to Jackson's, were based on the pro rata interest in their respective Bashir interviews ...

> **ii. How many records would Princess Diana have sold worldwide?**
>
> a) 157 million
> b) 335 million
> c) 680 million

Who might she have been? Britney? Christina? Anastasia?

Us Over There Them Over Here

Call the population of the USA 295 million.

Call the population of the UK 59.6 million.

Answers: i) a ii) b

British households spent an average of 26 pence each on commemorative flowers for Diana.

There are approximately 1.2 million Britons living permanently in the USA.

There are approximately 250,000 Americans living permanently in the UK.

i. What percentage of the US population is British?

a) 0.19
b) 0.41
c) 0.89

ii. What percentage of the UK population is American?

a) 0.42
b) 0.78
c) 1.88

Snap.

Hasta La Vista Baby

Currently 72,000 new homes are being built in Spain for Britons every year.

Assume every one of these homes represents real emigration from Britain to Spain (then pause a moment to consider why that doesn't seem such a ridiculous idea).

Forget about the other Britons buying homes which aren't new.

Britain's population of, say, 60 million is distributed throughout 24.83 million households, which suggests a Briton's home (wherever it may be) contains an average of 2.42 people.

Answers: i) b   ii) a

London enjoys 23 per cent more sunshine than Paris.

i. On the numbers above, at least how many people are theoretically emigrating to Spain every year?

a) 644,000
b) 365,000
c) 174,000

ii. How many years at this rate will it take for a million people to leave Britain for Spain?

a) 10 years 2 months
b) 5 years 9 months
c) 18 years 4 months

- *Approximate number of people emigrating from Spain to Britain every year: 15,000*

Lovely Greens

The World Wide Fund for Nature estimates that 200 new golf courses are either being planned or currently under construction around the Mediterranean coast.

A golf course uses the same amount of fresh water as a town of 12,000 people.

i. How many people could the fresh water required by 200 golf courses supply?

a) 3,200,000
b) 2,400,000
c) 4,800,000

Answers: i) c ii) b i) b

One in 35 people is now a migrant.

In Spain's parched south-east developers are reportedly planning to build at least 89 more golf courses. In Greece there are plans to build 40 and Cyprus is building eight.

---

ii. How many Cypriots could the fresh water required by eight golf courses supply?

a) 96,000
b) 146,000
c) 196,000

• *Population of Cyprus: 900,000*

Drinks in the clubhouse anyone?

Adam's Ale

On average you will drink 30,280 litres of water in a lifetime. An average bath holds 500 litres of water.

---

i. How many bathfuls of water will you drink in a lifetime?

a) 375.44
b) 60.56
c) 27.54

Answers: ii) a i) b

A golf course uses about 120 acres of land.

An elephant's trunk holds 15 litres of water.

ii. How many elephant trunkfuls of water will you drink in a lifetime?

a) 2,018.7
b) 1,456.2
c) 9,249.2

It wouldn't be much of a choice: dirty bath water or water from an elephant's nose ...

Suffer ~~the~~ Little Children

According to Oxfam, 1,000 people die each hour from hunger – most of them before their fifth birthday. In addition, 1,200 children under five die each hour from preventable disease. In addition, a child dies every fifteen seconds from lack of safe water and sanitation. The total number of deaths worldwide, therefore, every hour, from preventable disease, hunger and poor sanitation, is 2,440: a global 9/11 every hour. (Or almost: 2,749 died when terrorists destroyed the World Trade Center.)

i. If 2,440 people die every hour, how many die in a year?

a) 18,445,030
b) 21,389,040
c) 33,367,020

- *Total military and civilian fatalities in the First World War: 15,142,000*

Answers: ii) a i) b

30,280 litres of water in a lifetime works out at about a pint and a half a day.

> ii. How long does it take for 15,142,000 people to die from preventable disease, hunger and poor sanitation?
>
> a) About fourteen months (14.884)
> b) About eight and a half months (8.495)
> c) About five and three-quarter months (5.756)

So in the time it takes to make a new baby, more people (mainly children) die worldwide from preventable disease, hunger and poor sanitation than were killed in the First World War.

- *Infant mortality among children aged under one accounts for about 14 per cent of all deaths worldwide.*
- *In Sub-Saharan Africa, the Near East and North Africa infant mortality accounts for 20 to 25 per cent of all deaths.*
- *In North America, Europe, Japan, Australia and New Zealand infant mortality accounts for only 1 per cent of all deaths.*

It's an Old Plan, But It Might Just Work

You just can't trust some people: after a lifetime working long hours and raising their children, they refuse to invest what little spare cash they have in overpriced pension plans. It's an outrage. What ingratitude! How dare people prefer to spend their cash on enjoying themselves while they've still got the use of their legs! It's a scandal. Or rather, according to the British government, it's a crisis.

Economic surveys in the UK now point to a £57 billion ($102.4 billion) hole in the funds people in the UK will need to set aside to pay for their retirement. To

Answer: ii) b
Smoking kills 5 million people a year.

put this in perspective, to date the Eurofighter project has cost £55 billion ($98.8 billion).

To be in the top 1 per cent of British earners, you would need, at the time of writing, to earn at least £92,000 ($165,000) before tax. But £92,000 is peanuts to the very rich. Here is a list of the UK's 50 richest people (at the time this book was written).

1 Roman Abramovich £7.5bn	23= Sir Adrian and John Swire £1.2bn
2 Duke of Westminster £5bn	23= Viscount Portman and family £1.2bn
3 Hans Rausing £4.9bn	28= John Fredriksen £1.1bn
4 Philip Green £3.6bn	28= Roddie Fleming and family £1.1bn
5 Lakshmi Mittal £3.5bn	28= Sir Terry Matthews £1.1bn
6= Kirsten and Jorn Rausing £2.6bn	31 Robert Miller £995m
6= Sir Richard Branson £2.6bn	32 Roger and Peter De Haan £958m
8= Bernie and Slavica Ecclestone £2.3bn	33 Paul Fentener van Vlissingen £940m
8= Charlene and Michel de Carvalho £2.3bn	34 Nadhmi Auchi £898m
10 David and Simon Reuben £2.2bn	35 Urs Schwarzenbach £847m
11= Spiro Latsis and family £2.1bn	36= James Dyson £800m
11= Sri and Gopi Hinduja £2.1bn	36= Mark Pears and family £800m
13 Mahdi al-Tajir £2bn	38 Viscount Rothermere and family £798m
14= Boris Berezovsky £1.8bn	39 Sean Quinn and family £771m
14= Joseph Lewis £1.8bn	40 Sir Paul McCartney £760m
16 Lord Sainsbury and family £1.7bn	41 Lord Ashcroft £754m
17 Earl Cadogan and family £1.6bn	42= Lord and Edmund Vestey £750m
18 Eddie and Malcolm Healey £1.4bn	42= Sir David and Sir Frederick Barclay £750m
19= John Caudwell £1.3bn	44 Earl of Iveagh and Guinness family £725m
19= M. Czernin and H. de Walden family £1.3bn	45 Sir Alan Sugar £703m
19= Philippe Foriel-Destezet £1.3bn	46= David Bromilow £700m
19= Sir Ken Morrison and family £1.3bn	46= Michael Cornish and family £700m
23= Bruno Schroder and family £1.2bn	46= Richard Desmond £700m
23= Clive Calder £1.2bn	46= Sir Anthony Bamford and family £700m
23= Lady Grantchester and Moores family £1.2bn	46= Trevor Hemmings £700m

Source: *Sunday Times Rich List*

i. The combined wealth of how many of the UK's richest people could fill this £57 billion shortfall?

a) 34
b) 23
c) 14

Presumably all we'll need to do is ask.

Debt Cultures

The rich have an acronym: HNWIs (High Net Worth Individuals). Countries carrying the heaviest national debts have an acronym too: HIPC (Heavily Indebted Poor Countries). HIPC debt is what is often called Third World Debt. When activists demand 'Cancel the Debt', this is the debt they mean. Yet in the grand scheme of the global economy Third World Debt is fairly small. In total it comes to 'only' $263.7 billion (£146.8 billion). Probably more than you earn in a fortnight, but less than 0.75 per cent of annual global GDP ($36.36 trillion) and less than 2.5 per cent of the annual US GDP ($10.88 trillion).

Official explanations of how HIPC first accrued this debt might cite corruption, financial mismanagement and failed investment. They might even acknowledge the misfortunes of drought, epidemic, famine and war. But only rarely will they point to the self-interest of expatriate speculation, and they will dismiss as history any legacy of rapacious colonial imperialism by Old Europeans. Which is lucky for Old Europeans. Otherwise London, Madrid, Lisbon and Paris might have to share the blame.

Here is a list of Europe's 50 richest people (at the time of writing). Ten are British.

Answer: i) b

Thirteen per cent of the UK's national income is earned by 1 per cent of the population.

1	Karl and Theo Albrecht (Germany) $40.1bn	26=	Hansjorg Wyss (Switzerland) $6.1bn
2	Johanna Quandt (Germany) $19.4bn	26=	Landolt family (Switzerland) $6.1bn
3	Liliane Bettencourt (France) $18.3bn	28	Maria-Elizabeth and George Schaeffler (Germany) $5.9bn
4	Ingvar Kamprad (Sweden) $18bn	29	Curt Engelhorn (Germany) $5.8bn
5	Mikhail Khodorkovsky (Russia) $14.6bn	30	Friedrich Flick (Germany) $5.6bn
6	Roman Abramovich (UK) $13.5bn	31=	Alain and Gerard Wertheimer (France) $5.4bn
7	Brenninkmeyer family (Holland) $12bn	31=	Mikhail Fridman (Russia) $5.4bn
8	Bernard Arnault (France) $11.9bn	31=	Prince Hans-Adam (Liechtenstein) $5.4bn
9	Silvio Berlusconi (Italy) $9.7bn	34=	Adolf Merckle (Germany) $5.2bn
10=	Birgit Rausing (Sweden) $9bn	34=	Hasso Plattner (Germany) $5.2bn
10=	The Duke of Westminster (UK) $9bn	36	Antonia Johnson (Sweden) $5bn
12=	Amancio Ortega (Spain) $8.8bn	37	Maersk Mc-Kinney Moller (Denmark) $4.9bn
12=	Hans Rausing (UK) $8.8bn	38=	Héritiers Ströher (Switzerland) $4.7bn
14=	Oeri/Hoffmann family (Switzerland) $8.4bn	38=	Kirsten and Jorn Rausing (UK) $4.7bn
14=	Stefan Persson (Sweden) $8.4bn	38=	Mikhail Prokhorov (Russia) $4.7bn
16=	Ernesto Bertarelli (Switzerland) $7.7bn	38=	Sir Richard Branson (UK) $4.7bn
16=	Michael Otto (Germany) $7.7bn	38=	Vladimir Potanin (Russia) $4.7bn
18	Rudolf Oetker (Germany) $7.2bn	43=	Karl-Heinz Kipp (Germany) $4.5bn
19=	August von Finck (Germany) $6.7bn	43=	Francois Pinault (France) $4.5bn
19=	Fentener van Vlissingen family (Holland/UK) $6.7bn	45=	Luciano Benetton (Italy) $4.3bn
19=	Leonardo Del Vecchio (Italy) $6.7bn	45=	Erivan Haub (Germany) $4.3bn
22	Philip Green (UK) $6.5bn	45=	Reinhold Würth (Germany) $4.3bn
23=	Lakshmi Mittal (UK) $6.3bn	48=	Bernie and Slavica Ecclestone (UK) $4.1bn
23=	Serge Dassault (France) $6.3bn	48=	Charlene and Michel de Carvalho (UK) $4.1bn
23=	Walter Haefner (Switzerland) $6.3bn	50=	David and Simon Reuben (UK) $4bn

Source: *Sunday Times Rich List*

i. Combining the wealth of how many of Europe's richest would clear 'The Debt'?

a) 41
b) 23
c) 34

Answer: i) b

The nine richest people in Europe (excluding the UK) are worth more than the 50 richest in the UK.

ii. Excluding the UK-based (too busy paying off the UK pensions gap), the combined wealth of how many of Europe's other richest people could improve the lives of billions thus?

a) 19
b) 26
c) 23

No doubt their cheques are already written.

Never Play Cards With a Man Named 'Doc'

Currently, global earnings by the online poker industry are estimated to be $11 billion per annum, and the UK's biggest site has reported a rise in its number of registered players from 10,000 to 25,000 in the last year.

i. At this rate of growth, how long would it be before all 35.5 million 18–64 year olds in the UK were registered to play poker online?

a) A little over eleven years
b) A little under seven years
c) About eighteen years

Some card combinations in poker have nicknames. For example, a Dead Man's Hand is a pair of aces and a pair of eights, all clubs and spades. Why? Because reputedly Wild Bill Hickok was holding this hand when he was shot. (His final card was a two of spades.) This hand has appeared as a symbol of impending doom in various movies – like the western classic *Stagecoach*. These are modern examples.

Answers: ii) b   i) b

The world lottery market is worth $126 billion in sales.

A-A: American Airlines
K-J: Kojak
J-A: Jack Ass
J-J: Fishhooks
J-4: Flat tyre ('What's a jack for?')
10-4: Broderick Crawford, Over and Out
9-5: Dolly Parton
8-8: Snowmen
7-7: Walking Sticks, Sunset Strip

6-9: Big Lick
6-6: Route 66
5-10: Woolworth
4-5: Jesse James Colt 45
3-8: Raquel Welch
3-3: Crabs
2-2: Ducks
A-K-4-7: Assault Rifle
J-J-5-5: Motown (Jacks and Fives, Jackson Five)

Source: *Wikipedia*

How to win at poker? Here is a 'Cut Out and Keep' list of poker odds. The higher the odds against you getting the hand, the better your chance of it winning.

	Poker Odds			
	5 card		**7 card**	
	Odds you'll get this hand	Odds your opponent will get the same or better	Odds you'll get this hand	Odds your opponent will get the same or better
Your Hand	(1 in x)	(1 in x)	(1 in x)	(1 in x)
Nothing	2.00	1.00	5.74	1.00
One pair	2.37	2.00	2.28	1.21
Two pair	21.04	13.11	4.26	2.58
Three of a kind	47.33	34.83	20.70	6.55
Straight	254.80	131.82	21.65	9.58
Flush	508.80	273.11	33.05	17.18
Full house	694.17	589.60	38.52	35.77
Four of a kind	4,165.00	3,914.10	595.00	502.13
Straight flush	72,193.33	64,974.00	3,590.57	3,217.21
Royal flush	649,740.00	649,740.00	30,940.00	30,940.00

Insomnia

On average we stay awake for 16.9 hours a day.

People in the UK, at the time of writing, earn an average of £25,170 ($45,220) a year each.

> i. If a 16.9-hour day earns £25,170, then if we never went to sleep how much should we earn, on average, per year, theoretically?
>
> a) £92,655.44 ($166,464.76)
> b) £53,765.98 ($96,595.96)
> c) £35,744.38 ($64,218.35)

Now stick your own salary into this equation. See how much you would earn. It's not reality, it's Number Freaking.

Rat Race at a Snail's Pace

America's car fetish has substantially reduced how far Americans walk. So while filming *Super Size Me*, his 2004 'diary' of what a month eating at McDonald's did to him, film director Morgan Spurlock set himself a daily 'walking limit' of 2,000 steps – the American average …

- *Doctors say for a healthy cardiovascular system you should walk 10,000 steps daily.*

At a step length of half a metre (for argument's sake) 2,000 steps is 1 kilometre (and 10,000 steps is 5 kilometres).

The average American is awake for 16.9 hours a day.

Answer: i) c
On average we dream for about 100 minutes a night.

i. Spreading 1 kilometre evenly across their waking day, the average American can be said to be moving at what average speed (under their own steam)?

a) 118 metres per hour
b) 59 metres per hour
c) 19.5 metres per hour

According to the *Guinness Book of Answers* ...

- *The common house snail moves at a speed of 49 metres per hour.*
- *The three-toed sloth moves at a speed of 109 metres per hour.*
- *The giant tortoise moves at a speed of 370 metres per hour.*

Thus a tortoise moves over six times faster than the average American. And a sloth nearly twice as fast.

Perhaps the amazing thing isn't that the first Americans ever tamed the Wild Frontier, but that they ever got there at all.

How Far Is the Wild Frontier Anyway?

At 2,000 steps per day, over a lifetime you'd walk a surprisingly long way.

i. How many steps is 2,000 steps per day in a year?

a) 1,300,600
b) 730,500
c) 467,000

Answers: i) b   i) b

A six-foot-tall sprinter has a stride length of about 2.5 metres.

The life expectancy of an average American (combining the numbers for men and women) is 77 years. Assuming they can walk 2,000 steps a day from at least the age of four, that's 73 walking years.

(When talking of averages it doesn't matter that some Americans are babies, elderly or infirm. The average American is always average age, height, health and wealth, and stays that way for the entire duration of the average life span.)

ii. So, theoretically, how many steps is that in a lifetime?

a) 135,326,500
b) 53,326,500
c) 23,198,200

iii. So assuming that average step length of, say, half a metre, how far does the average American walk in a lifetime?

a) 26,663.25 kilometres (16,568.23 miles)
b) 60,223.24 kilometres (37,422.01 miles)
c) 11,456.33 kilometres (7,118.83 miles)

This is very roughly the same as strolling from the top of North America to the bottom of South America and back again. (You know what it's like, you pop out to the shops, realise you've forgotten your wallet, and have to go all the way back to get it.)

Alternatively it's very approximately ...

- *eight times the distance from Toronto to Vancouver (3,366 kilometres)*
- *seven times the distance from Sydney to Perth (4,100 kilometres)*

Answers: ii) b iii) a

Thirty-eight per cent of North America is wilderness.

- *six times the distance from New York to Los Angeles (4,544 kilometres)*
- *five times the distance from New Delhi to Tokyo (5,839 kilometres)*
- *four times the distance from Tel Aviv to Cape Town (7,486 kilometres)*
- *three times the distance from London to Singapore (10,846 kilometres)*

Any which way it's easily far enough to pioneer a route to the Wild Frontier ...

Priced by the Yard

Tiger Woods was born on Tuesday 30 December 1975. So prodigious was his golfing talent that in 1978, while still a toddler, he appeared on national television swapping driving tips with Bob Hope. Despite this, he is now reported to have an income of $80 million per year.

Notwithstanding the proportion of these millions he harvests from sponsor-ship fees, it is fair to say he earns *all* his money from his ability to play golf brilliantly.

There is no such thing as a standard golf course, but there are guidelines.

USGA Yardage Guidelines for Golf Course Holes (Men):

Par 3: up to 250 yards

Par 4: 251–470 yards (average = 360 yards)

Par 5: 471–690 yards (average = 580 yards)

Par 6: over 690 yards

Let us assume that to maintain his game Tiger Woods plays, on average, whether in competition or practice, eighteen holes of golf each and every day, and that to cover all the bases he strives to play an even mix of par 4 and par 5 holes. The average guideline length of a par 4 or 5 hole is 470 yards (half the sum of 360 + 580). Therefore, we shall assume Tiger plays 18 x 470 (8,460) yards of golf a day, or 3,090,015 yards (2,825,510 metres) annually. Finally let us assume Tiger has an average stride length of 0.5 metres.

Assuming these estimates are correct (and since we're Number Freaking, of course they are), we can calculate that Tiger Woods treads 5,651,020 steps on golf courses every year.

So if Tiger's wealth ultimately stems from playing the game ...

---

i. How much does Tiger earn (of $80 million) for each and every step (of 5,651,020) he takes on a golf course?

a) $17.70 (£9.85)
b) $14.16 (£7.88)
c) $35.40 (£19.70)

So if an American walking the national average of 2,000 steps a day (rather than the 10,000 that doctors recommend) was paid as well as Tiger, he or she would earn $28,320 a day.

And as to Tiger and his amazing skill? If it's any consolation there are no reports he's any good at table tennis.

A Free Fitting with Every Stop and Search?

London's Metropolitan Police has announced plans to merchandise its logo ...

For an official 'I've been nicked by the Met' souvenir tee-shirt perhaps? Other police forces will presumably follow suit. Think: 'I'm with Somerset and Avon

---

Answer: i) b

Michael Schumacher is the only sportsman believed to earn as much as Tiger Woods.

Armed Response Unit', or 'Merseyside CID Vice Squad do it in manacles', or 'My boyfriend was arrested by West Midlands Serious Crime Squad and all I got was this lousy tee-shirt' ...

There are a grand total of 151,678 police officers in the UK. How many other retailers can muster a staff like that? They could swear an oath of allegiance to the Queen – and always to stock a full range of colours and sizes.

The land area of the UK excluding reservoirs and lakes is 241,590 square kilometres.

i. How big a sales area could each police officer be responsible for?

a) 1.59 square kilometres
b) 2.17 square kilometres
c) 4.56 square kilometres

The total UK population is 59.66 million people.

ii. How many potential customers could each police officer be responsible for?

a) 224
b) 393
c) 523

Since police officers work shifts and patrol in pairs, the average area that each pair is actually responsible for on each shift is 9.54 square kilometres. Assuming every officer worked the beat, on average you're always within about 1,500 yards of a police patrol in the UK.

Answers: i) a ii) b

Sixty-five per cent of children with a father in jail end up in prison.

By contrast the UK's custodial prison population numbers about 74,000. That's a pair of police officers for every prisoner. But what if these prisoners escaped and dispersed equally? There'd be one prisoner per 3.26 square kilometres. And on average, one would be closer than a cop – within 1,000 yards of where you are now.

The Met has 30,265 serving police officers, plus 750 voluntary constables, 1,430 community support officers and 500 traffic wardens – in other words a total of 32,945 beat officials with which to serve London's 7.17 million inhabitants. It has an annual budget of £2.33 billion ($4.19 billion). The shortfall it hopes merchandising can replace is £17 million ($31 million).

---

iii. How many potential customers could each uniformed Met official have?

a) 842
b) 429
c) 218

---

iv. How often must every uniformed official of the Metropolitan Police sell a £10 tee-shirt to make up the shortfall?

a) Once a month
b) Once a week
c) Once a day

At the time of writing it's still illegal to impersonate a police officer – even if they sell you the costume.

Answers: iii) c iv) b

One in twelve of all Britain's workforce is a cleaner.

Some Useful United States Police Radio Codes

10-1	Call your command	10-20	Robbery has occurred
10-2	Return to your command	10-21	Burglary has occurred
10-3	Call dispatcher by telephone	10-24	Assault has occurred
10-4	Gotcha, over and out	10-30	Robbery in progress
10-5	Repeat message	10-31	Burglary in progress
10-6	Standby	10-33	Explosive device or threat
10-7	Verify address	10-34	Assault in progress
10-10	Possible crime (prowler, shots fired, suspicious person/vehicle)	10-29	Other crime (specify) has occurred
10-11	Alarm (specify type)	10-39	Other crime (specify) in progress
10-12	Police officer holding suspect	10-50	Disorderly person/group or noise
10-13	Assist police officer	10-51	Roving band (specify direction of travel and number in group)
10-14	Licence plate check (occupied and suspicious – verify if stolen)	10-52	Dispute (specify)
10-15	Licence plate check (verify if stolen – occupied or not)	10-53	Vehicle accident (specify)
10-16	Vehicle is reported stolen	10-54	Ambulance case (specify type)
10-17	Vehicle is not reported stolen	10-62	Out of service – mechanical (give condition and location)
10-18	Warrant check shows an active warrant	10-63	Out of service – meal break
10-19	Warrant check negative	10-66	Unusual incident (train derailment/collision, plane crash, building collapse)
		10-84	Arrived at scene

1-8-7 is the code for murder used by police officers, particularly in the state of California, because it is Section 187 of the California Penal Code that deals with the crime.

One in ten arrests in America is for drunk driving.

Bullet Points

The Intifada cost Israel 4 per cent of GDP – an annual cost to the average Israeli of $788.

i. How much did the Intifada cost the average Israeli every day?

a) $21.90
b) $13.76
c) $2.16

But it's still safer than being in DC.

Annual *187* rates per 100,000 of population (the hip-hop slang for murder is also '187')

USA

New Orleans	100.0	USA (men aged 15–29)	23.6	Los Angeles	17.0
Washington DC	46.0	Chicago	22.0	Boston	10.0
New York	45.0	Philadelphia	19.0	USA (general)	6.9

In a survey in four US cities, homicides were described as 'drug related' in 25.7 to 44 per cent of cases. The most common feature of homicide is alcohol.

Rest of World

Colombia	61.6	Venezuela	16.0	New Zealand	1.5
El Salvador	55.6	Mexico	15.9	Canada	1.4
Jamaica	44.0	Bahamas	14.9	Italy	1.1
South Africa	42.7	Estonia	14.8	Germany	0.9
Brazil	23.0	Belarus	10.5	Netherlands	0.8
Russian Federation	21.6	Israel	4.0	UK	0.8
Iraq	21.0	(non-terrorism 0.5; terrorism 3.5)		France	0.7
(non-terrorism 6; terrorism 15)		India	3.0	Japan	0.6
Albania	21.0	Australia	1.6	Spain	0.8

Sources: various

Answer: i) c

The weight of 2.5 AK47 bullets is about 21 grams.

By the Time You Finish Reading This Someone Will Be Dead

IANSA - the International Action Network on Small Arms - estimates that 16 billion bullets are produced every year.

i. Based on a global population of 6.388 billion, how many bullets is that each?

a) 1.3
b) 2.5
c) 2.9

It's said that during the Vietnam War, US forces fired 50,000 bullets per enemy killed.

ii. If it takes 50,000 per 'kill', about how many people should be shot to death a year?

a) 448,000
b) 760,000
c) 320,000

In fact according to the World Health Organisation, there are 558,000 homicides worldwide every year. Obviously not all murders involve guns.

Answers: i) b ii) c

There are 100 million unexploded landmines on earth.

iii. So how often is someone somewhere murdered?

a) Every 14.98 seconds
b) Every 56.55 seconds
c) Every 166.22 seconds

And how long is that? Read this number freak aloud, and by the time you get to here, someone else will have been murdered.

Saturday Night Special

i. If the world's 558,000 murders were evenly dispersed across its 148,327,070 square kilometres of land, on average one murder per year would occur in how big an area?

a) 250.34 square kilometres
b) 289.41 square kilometres
c) 265.82 square kilometres

- *Combined area of the Cayman Islands: 259 square kilometres*

The WHO world report on violence and health 2002 cites a 1992 study that estimated the annual cost of treating gunshot wounds in the USA at $126 billion. At current prices that's $167.8 billion.

ii. Given a US population of 295 million, how much is this per year per American?

a) $1,157
b) $122
c) $569

Answers: iii b i) c ii) c

Sixty per cent of the world's 639 million small arms are in civilian hands.

iii. And how much is this, per American, each and every day?

a) $1.45
b) $1.66
c) $1.56

- *Gasoline, average US price per gallon, 2003: $1.56*
- *Average daily mileage cost to Americans at 20 miles per gallon of gasoline: 20*

Little and Often

Collectively Americans undertake over 4 trillion miles in personal travel every year – an average of around 15,000 miles of personal travel per head per year.

i. How far does every American travel, on average, each and every day?

a) 46.22 miles
b) 41.07 miles
c) 42.89 miles

- *Distance Los Angeles to Laguna Beach, California: 45 miles*
- *Distance Maidstone to Dover, UK: 41 miles*
- *Distance Tel Aviv to Jerusalem: 41 miles*

Answers: iii) c i) b

An infantry battalion can march 40 miles in a day.

Stating the Obvious

Sudan is roughly the size of Western Europe.

- *Twenty-eight per cent of Africa is wilderness.*

In rural Sudan a little thing called abject poverty obliges millions of women to spend several hours every day collecting water. On foot.

Assume this involves a daily round trip of at least 5 kilometres.

i. How far will a Sudanese woman walk collecting water if she treks 5 kilometres every day for 40 years?

a) 132,093 kilometres
b) 73,050 kilometres
c) 23,098 kilometres

ii. The circumference of the earth is 40,075 kilometres. How many times around the world will a Sudanese woman trek the equivalent of, during her 40-year quest for water?

a) 7.33
b) 3.96
c) 1.82

- *Life expectancy in Sudan is 58. In America it is 77.*

It's still better to be rich and drive, than poor and walk ...

Answers: i) b   ii) c

America has an infant mortality rate of 6.9 per 1,000. In Sudan it is 70 per 1,000.

Ethnocentric

Americans believe, apparently, that, on average, the ideal age is 41.

Obviously some folks would disagree.

Eight Countries Where Life Expectancy Is Less Than 41

Country	Life Expectancy	Population
Mozambique	40.87	18,811,731
Zambia	39.36	10,462,436
Zimbabwe	37.82	12,671,860
Swaziland	37.54	1,169,241
Malawi	37.48	11,906,855
Lesotho	36.81	1,865,040
Angola	36.79	10,978,552
Botswana	34.19	1,561,973

i. How many people, based on the list above, would definitely disagree with the proposition that the ideal age is 41?

a) 69,427,688
b) 18,811,731
c) zero – 'cos if you get there ...

- In 1800, life expectancy in France was just under 30; in Britain it was about 36.

Answer: i) a

At age 41 the average adult on earth has only lived about 60 per cent of their life.

Aid for Aids

Worldwide, 37.2 million adults now suffer from HIV/Aids. In 2003 there were 4.9 million new infections and 3.1 million Aids-related deaths. Currently there are 14,000 new cases and 8,500 deaths daily. Thankfully anti-retroviral drugs (ARVs) to treat adult Aids can now be had for as little as $140 a year, but tragically, only one in ten sufferers in the developing world can get the drugs they need. People taking ARVs must do so for life.

i. How much would it cost to provide these ARVs for one year to all adults with Aids?

a) $4,894,000,000 (£2,724,034,200)
b) $5,208,000,000 (£2,898,808,800)
c) $9,332,000,000 (£5,194,255,800)

Sadly 2.2 million children in the world also have HIV/Aids. They can't be treated with adult drugs. Their ARV drugs cost $1,500 a year – and must be taken for life too.

ii. How much would it cost to provide ARVs for one year to all children with HIV/Aids?

a) $1,453,000,000 (£808,749,860)
b) $3,300,000,000 (£1,836,802,800)
c) $5,600,000,000 (£3,116,998,700)

The combined total is $8,508,000,000 (£4,735,611,700).

Here then is a list of the world's 50 richest individuals and families at the time of writing – when their combined net worth was $813.2 billion (£452.6 billion).

Answers: i) b   ii) b

On average, untreated HIV causes Aids after about ten years and death two years later.

1 Robson Walton and Family (USA) $97.4bn	25= Li Ka-shing (Hong Kong) $12bn
2 Bill Gates (USA) $45.5bn*	25= Steve Ballmer (USA) $12bn
3 Warren Buffett (USA) $41.9bn	28 Bernard Arnault (France) $11.9bn
4 Karl and Theo Albrecht (Germany) $40.1bn	29 The Kwok brothers (Hong Kong) $11.1bn
5 Forrest Jr Mars and Family (USA) $30.4bn	30 John Kluge (USA) $10.2bn
6 King Fahd (Saudi Arabia) $24.3bn	31= Sheikh Makhtoum (UAE) $9.7bn
7 Barbara Cox Anthony and Anne Cox Chambers (USA) $21.7bn	31= Silvio Berlusconi (Italy) $9.7bn
8 Prince Alwaleed (Saudi Arabia) $20.8bn	33= Birgit Rausing (Sweden) $9bn
9 Paul Allen (USA) $20.5bn	33= Duke of Westminster (UK) $9bn
10= Johanna Quandt (Germany) $19.4bn	35= Amancio Ortega (Spain) $8.8bn
10= Sheikh of Abu Dhabi (Abu Dhabi) $19.4bn	35= Charles Ergen (USA) $8.8bn
12 Liliane Bettencourt (France) $18.3bn	35= Hans Rausing (UK) $8.8bn
13 Larry Ellison (USA) $18.2bn	38 Sumner Redstone (USA) $8.6bn
14 Ingvar Kamprad (Sweden) $18bn	39= Oeri/Hoffmann family (Switzerland) $8.4bn
15 Kenneth Thomson (Canada) $16.7bn	39= Stefan Persson (Sweden) $8.4bn
16 Samuel and Donald Newhouse (USA) $14.9bn	41 Pierre Omidyar (USA) $8.3bn
17 Robert and Thomas Pritzker (USA) $14.7bn	42 Nasser al-Kharafi (Kuwait) $8.1bn
18= Emir of Kuwait (Kuwait) $14.6bn	43= Ernesto Bertarelli (Switzerland) $7.7bn
18= Mikhail Khodorkovsky (Russia) $14.6bn	43= Michael Otto (Germany) $7.7bn
20 Abigail and Edward Johnson (USA) $14.4bn	45= Galen Weston (Canada) $7.6bn
21 Sultan of Brunei (Brunei) $13.8bn	45= Rupert Murdoch (USA) $7.6bn
22= Carlos Slim Helu (Mexico) $13.5bn	47 Carl Icahn (USA) $7.4bn
22= Roman Abramovich (UK) $13.5bn	48= Rudolf Oetker (Germany) $7.2bn
24 Michael Dell (USA) $12.6bn	48= Samuel Johnson (USA) $7.2bn
25= Brenninkmeyer family (Holland) $12bn	50 Philip Knight (USA) $6.8bn

Source: *Forbes*

* When list was compiled

iii. How many years could the world's 50 richest people supply the entire current global population of Aids sufferers with ARV therapy?

a) 77.8 years
b) 95.6 years
c) 66.2 years

The world's richest 50 people could give a full lifetime back to every Aids sufferer on earth.

Answer: iii) b

Worldwide, 8.8 million people contract TB every year.

How Much Does the Queen Cost?

Assume average UK household income is the same as the average UK wage (£25,170).

Now imagine £25,170 looks like this: £ (1)

By this measure the US President, who gets $400,000, gets this: £££££££££ (9)

Or, more accurately, this: $$$$$$$$$$$$$$$$ (16)

Meanwhile the Queen's household gets this (£36,392,600):

££
££
££
££
££
££
££
££
££
££
££
££
££
££
££
££
££
££
££
££
££

££
££££££££££££££££££££££ (1,446)

A Mere Trifle

In June 2004 Buckingham Palace claimed that running the British royal household in 2003 cost the British population the equivalent of 61 pence (£0.61) per head per year. A price, said the palace, equivalent to two pints of milk each. (For 2002 they said it was a loaf of bread.)

The accredited population of Britain (at the time of writing) is 59.66 million people.

---

i. At 61 pence each, how much is this overall?

a) £127,879,341 ($229,748,024)
b) £61,453,990 ($110,408,238)
c) £36,392,600 ($65,382,945)

---

ii. Two pints is 1.137 litres. So, based on a figure for the UK population of 59.66 million, what is the total cost of running the royal family in milk?

a) 135.85 million litres
b) 67.83 million litres
c) 178.22 million litres

For the sake of comparison the US President receives a basic salary of $400,000 (£222,643).

Answers: i) c ii) b

Dividing their wealth equally, America's 25 richest people (average age 60.6) could give everyone on earth $32.33.

iii. At UK prices, how much milk is the US President's basic salary equivalent to?

a) 1,593,234 litres
b) 863,119 litres
c) 414,992 litres

Full cream or semi-skimmed, that's a lot of trifle.

Cowpie

There are estimated to be 1,294,604,000 head of cattle on earth. There are 6.388 billion people.

i. How many cattle are there per person on earth?

a) 0.203 (4.93 people per cow)
b) 0.434 (2.30 people per cow)
c) 0.320 (3.13 people per cow)

There is no rule that says all these steers couldn't potentially be eaten. There is no rule that says all this meat couldn't be turned into quarterpounders. An 850-pound beef carcass yields about 310 pounds of edible meat.

ii. Globally, how many quarterpounders are masquerading as cows right now?

a) 2,134,786,300,000 (2.13 trillion)
b) 1,605,308,900,000 (1.61 trillion)
c) 6,004,003,800,000 (6.00 trillion)

Answers: iii) c i) a ii) b

Twenty-one per cent of young Americans would like to be US President some day.

Eaten as a mono-food these patties could provide all the 4.472 billion adults on earth with their maximum recommended calorie allowance for about 36 days ... Or, put another way, all the cattle on earth could feed adult humanity for about five weeks.

Cowadunga

The world's 1,294,604,000 cattle each poop up to sixteen times and produce 65 pounds (29.48 kilograms) of manure daily.

i. How many metric tonnes of cattle poo are produced every day?

a) 38,164,926
b) 11,747,273
c) 27,857,089

ii. How many metric tonnes of cattle poo are produced every year?

a) 13,939,739,000 tonnes
b) 39,234,234,000 tonnes
c) 22,456,050,000 tonnes

At one cubic metre per metric tonne, in one year this much cow manure could cover an area about two and a half times the size of Rhode Island to the height of a man, or almost fill Lake Lucerne.

Answers: i) a ii) a

At 15 cents each, buns for 1.61 trillion hamburgers would cost about $240 billion (220 days' US military spending).

iii. How many tonnes of cow pat is this per person worldwide per year?

a) 4.56 tonnes
b) 0.45 tonnes
c) 2.18 tonnes

Disposing of this much cow frisbee is an enormous – and serious – problem.

According to the census of 2001, as many as 25 million homes in India (roughly the total number of households in the UK) incorporate dung in their structure. Remarkably some of these homes have stood for over a century, and some are reputed to have survived for over 800 years.

As many as two-thirds of households in the developing world depend on dung as a significant fuel source. Tragically, as many as 4 million women and children are estimated to die every year from respiratory disease triggered by the smoke from wood and dung cooking fires.

Worldwide

For a sense of how fast the global population grows, according to the US Bureau of the Census, in one hour between 5.30 p.m. and 6.30 p.m. on 2 May 2005 the net growth in the global population was 8,470.

If we imagine the entire global population is a village of precisely 200 people, here are some things we could observe.

Answer: iii) c

Rhode Island covers 3,144 square kilometres. Lake Lucerne holds 14.5 billion cubic metres of water.

- 97 would be women (at birth)
- 103 would be men (at birth)
- 34 would be left-handed
- 60 would be under 15 years old
- 14 would be over 65 years old
- 38 would come from the more developed countries
- 162 would come from the less developed countries
- There would be 122 Asians
  (including 38 from China, 34 from India and 6 from Indonesia)
- There would be 24 Europeans
- There would be 28 Africans
  (including 22 who live in the sub-Sahara)
- There would be 18 from South America and the Caribbean
- There would be 10 from North America
  (including 9 Americans)
- There would be 1 from Oceania
- 120 would live within 62 miles of a coastline
- 96 would be urban dwellers
- 50 would be homeless or live in sub-standard housing
- 96 would lack access to basic sanitation
- 32 would lack access to safe drinking water
- 28 would suffer from malnutrition
- 32 would be unable to read and write
- 58 would believe in witchcraft
- 9 would get drunk each day
- 1 would eat at McDonald's each day

The global infant mortality rate is 55 per 1,000 births.

And It Stains Your Fingers

The average life expectancy in Britain is now 78 years.

A British smoker aged 45 has three times the chance of dying by age 60 than a non-smoker.

i. If a smoker smokes ten cigarettes a day between the ages of 15 and 60, how many cigarettes will they consume in a lifetime?

a) 179,456
b) 164,363
c) 34,483

ii. At £4.50 for a packet of twenty, how much would all these cigarettes cost?

a) £29,305.40 ($52,650.08)
b) £36,981.68 ($66,441.28)
c) £43,365.76 ($77,910.92)

- *Average annual UK salary (at the time of writing): £25,170*

Getting the Hell Out of Dodge

Persecuted by changing laws and lifestyles in the West, multinational tobacco companies increasingly turn their gaze towards the developing world. Their Holy Grail is China, where there are 320 million smokers.

Answers: i) b ii) b

In 2004, 106,000 people in the UK died from active smoking.

There are more smokers in China than any other country on earth. Sixty per cent of adult Chinese men smoke – but only 4 per cent of women.

> **i. Adult Chinese smokers smoke 1.7 trillion (1,700,000,000,000) cigarettes a year. So about how many cigarettes does a Chinese smoker smoke on average per day?**
>
> a) 14.5
> b) 36.3
> c) 9.7

(The 27 per cent of UK adults who smoke average 15–18 cigarettes per day.)

The price of a locally produced pack of twenty cigarettes – which the vast majority of Chinese smokers still prefer – is around £0.50 ($0.90).

> **ii. On this basis about how much do Chinese smokers spend on cigarettes each year?**
>
> a) £123.6 billion ($222.1 billion)
> b) £42.5 billion ($76.4 billion)
> c) £83.5 billion ($150.0 billion)

By comparison, British smokers spend about £15.5 billion and American smokers about $80.3 billion annually.

> **iii. Given that the retail price of cigarettes in the UK is around nine times the Chinese price, what would the Chinese cigarette market be worth at UK prices?**
>
> a) £123 billion ($221 billion)
> b) £382 billion ($686 billion)
> c) £564 billion ($1.01 trillion)

Answers: i) a ii) b iii) b

No wonder Western tobacco companies are so keen to plant their butts in China.

Burning Questions

i. Assume the average cigarette is 85 millimetres long. What is the combined total length of all the 1.7 trillion cigarettes currently smoked in China every year?

a) 167.4 million kilometres (104 million miles)
b) 144.5 million kilometres (90 million miles)
c) 133.6 million kilometres (83 million miles)

● *Distance from the earth to the sun: 150 million kilometres*

(You can make up your own joke about smokers getting a light!)

It's been estimated that the Chinese smoke a third of all the cigarettes consumed worldwide, numbering annual worldwide cigarette consumption at 5.1 trillion sticks.

ii. How many cigarettes do the world's smokers smoke every day?

a) 45,564,906,000
b) 13,963,039,000
c) 21,223,657,000

(Which is equivalent to about two a day for every one of the world's 6.388 billion people.)

Answers: i) b ii) b

There are 240 million smokers in India.

If we assume each cigarette has a volume of 5 cubic centimetres, 5.1 trillion would fill a cubic box with sides of 294 metres – just less than the height of the Eiffel Tower.

Laid end to end, as a single cigarette, 5.1 trillion cigarettes would stretch 269.4 million miles or 433.5 million kilometres – about the distance from the sun to Mars and back.

iii. So if the smokers of the world collectively consume a cigarette of this combined length every year, at what speed are the world's smokers consuming cigarettes?

a) 28,387 mph (45,675 kph)
b) 30,732 mph (49,457 kph)
c) 33,213 mph (53,440 kph)

- *Estimated speed of the meteorite triggering dinosaur extinction: 50,000 kph*

A Great Black Smudge

Each person in the UK – all 60 million people – now produces, apparently, more than 5 tonnes of carbon pollution a year, on average. These carbon emissions come in various chemical guises, and from a range of human activities: agriculture, industry, cars, domestic heating, etc., but wherever and however they might arise, they are all just some variant on coal.

Imagine each of us disposed of our emissions in a 5-tonne truck.

A 5-tonne coal truck is 6.93 metres long.

Answer: iii) b
Americans smoke about 450 billion cigarettes a year.

i. How far would a convoy of 60 million 5-tonne trucks laden with coal stretch, travelling nose to tail?

a) 350,087 kilometres
b) 415,800 kilometres
c) 675,000 kilometres

- Total length of UK roads (1999): 371,913 kilometres

Shiver Your Timbers

The UK gets through 12.6 million tonnes of paper and board every year. In Austria and Germany over 70 per cent of paper gets recycled. In the UK only 39 per cent gets used again.

i. How much extra paper could be conserved annually if the UK recycled an extra 31 per cent?

a) 2,345,000 tonnes
b) 5,556,000 tonnes
c) 3,906,000 tonnes

It takes seventeen trees to produce a tonne of paper.

ii. How many extra trees could be conserved annually if paper was recycled as diligently in the UK as in Austria and Germany?

a) 145,569,000
b) 66,402,000
c) 23,465,000

Answers: i) b i) c ii) b

The global average is about 1 tonne of carbon pollution per person per year.

Planted at 2 metre intervals this many trees could create a 266 square kilometre forest.

- *Area of the Cayman Islands: 259 square kilometres*

In 1805 Nelson won the Battle of Trafalgar from the bridge of HMS *Victory* with a fleet of 27 ships. HMS *Victory* was built from 2,500 oak trees.

iii. How many ships the size of HMS *Victory* could be built from all the extra trees that would be conserved annually if paper in the UK was recycled as diligently as in Austria?

a) 8,345
b) 26,561
c) 64,394

The Spanish Armada that sailed for England in 1588 consisted of 130 ships. Each of these galleons was built from over 2,000 pine, oak, mahogany and cedar trees.

iv. How many ships the size of a Spanish galleon could be built from the extra trees that would be conserved annually if paper was recycled as diligently in the UK as in Germany?

a) 33,201
b) 24,493
c) 17,345

Answers: iii) b iv) a

The world's heaviest giant sequoia weighs about 670 tonnes.

Typical

According to the CIA there are 268 nations, dependent areas, territories, other, and miscellaneous 'administrative districts' on earth. We'll think of them as countries.

Global population is 6.388 billion.

i. How many people are there on average per nation, dependent area, territory, other, or miscellaneous administrative district?

a) 23.84 million
b) 16.77 million
c) 30.04 million

- *Population of Malaysia: 23.52 million*
- *Area of Malaysia: 329,750 square kilometres*

The land area of the earth is 148,327,070 square kilometres.

ii. If all 268 administrative districts were the same size how much land would they each occupy?

a) 553,459 square kilometres
b) 467,761 square kilometres
c) 679,954 square kilometres

- *Total area of France: 547,030 square kilometres*
- *Population of France: 60.4 million*

So the world's most average country would be the size of France with the population of Malaysia.

Answers: i) a ii) a

The least densely populated country on earth is Mongolia with 1.66 people per square kilometre.

All the Beer You Can Drink

An average drinker in the UK has a potential boozing career of 60 years (The law says you can start drinking at 18, and average life expectancy is 78.)

According to the UK Office of National Statistics, on average a British adult in work drinks 16.2 units of alcohol a week. (Which roughly translates into the popular national pastime of one or two drinks a day for six days, and one day getting bombed.)

---

i. If half a pint of ordinary strength beer contains one unit of alcohol, how many gallons of beer (a gallon is 8 pints) would the average drinker consume in their lifetime if that was the only way they got their 16.2 units every week?

a) 1,754
b) 3,159
c) 5,674

---

ii. A standard domestic bath holds 110 gallons (0.5 cubic metres). How many bathfuls of beer is this?

a) 52.76
b) 28.72
c) 16.44

Answers: i) b ii) b

Thirty per cent of beer sold in Britain is ale. Seventy per cent is lager, stout or cider.

> ### iii. And if the beer cost £3 ($5.39) a pint (as it now does, pro-rata, in many a metropolitan bar), how much would all that beer cost in total?
>
> a) £160,882 ($289,041)
> b) £75,816 ($136,211)
> c) £334,155 ($600,343)

In the UK, based on an average salary of £25,170, a lifetime's beer will cost about three years' gross salary. Allowing for taxes and deductions, we'll call it four years. For a graduate with a working career of 44 years from age 21 to 65, this means a dedicated average beer drinker will spend at least 7 per cent of everything they ever earn on beer.

For the record, Bill Gates earns enough to pay for a lifetime's beer every 42 minutes.

Drinking and Driving

In 2001 (the last year for which figures are available) the total volume of beer consumed worldwide was 140 billion litres.

Litres of beer per head per annum	
Germany	120
Australia	109
UK	104
USA	83

Answer: iii) b
A barrel of beer is 36 gallons. A beer drinker would drink 87.75 in a lifetime.

The average flow of water over Niagara Falls is 6,000 cubic metres per second.

i. How long would it take all this beer to flow over Niagara Falls (in hours and minutes)?

a) 6 hours 29 minutes
b) 18 hours 18 minutes
c) 27 hours 4 minutes

Coniston Water, the British lake forever infamous for the fatal, 300 mph World Water Speed Record attempt by Donald Campbell in *Bluebird* in January 1967, holds 113.3 billion cubic metres. It has a surface area of 4.91 square kilometres.

ii. To what depth in metres could the world's beer production fill Coniston Water?

a) 44.788
b) 28.513
c) 16.392

You've heard of a wine lake? This is what the world's beer lake would look like FOR REAL.

In Vino Veritas

To make an average bottle of wine takes 600 grapes.

An average drinker in the UK consumes the equivalent of 8,424 bottles of wine in a lifetime.

Answers: i) a ii) b

In 1928 prohibition-era America, US doctors earned $40 million selling prescriptions for whisky.

i. How many grapes would such a dedicated UK wine drinker consume as wine in a lifetime?

a) 4,908,300
b) 5,054,400
c) 6,756,000

There are 75 grapes in a bunch. That's 67,392 bunches in a lifetime. A grapevine produces about 40 bunches of grapes. That's 1,685 grapevines in a lifetime.

A vineyard can support about 800 grapevines per acre.

ii. How many grapes is that per acre?

a) 13.3 million
b) 7.5 million
c) 2.4 million

That's 4,000 bottles per acre.

In a lifetime an average wine drinker will consume roughly the annual output of a two-acre vineyard. Most Bordeaux Appellation vineyards are less than 20 acres.

Ones and Twos

For those readers who might prefer to pass, here follows a short burst of Number Freaking around the topic of a limited range of human bodily waste products.

Answers: i) b   ii) c

In 1924, during prohibition, the US drank 2,944,700 gallons of communion wine.

Some sources record that an average healthy human adult passes 1,150–1,500 millilitres of urine a day. More exactly others report that we each make (and pass) the equivalent of 1 millilitre a minute.

i. At one millilitre a minute how much is this per day?

a) 1,440 millilitres a day
b) 1,380 millilitres a day
c) 1,260 millilitres a day

ii. Assuming everyone on earth were healthy (ha!) and the total world population over the age of 15 (that's 4.42 billion people) counts here as adult, how much urine does adult humanity produce a year?

a) 12,421,607,500,000 litres (12.42 billion cubic metres)
b) 5,856,565,700,000 litres (5.86 billion cubic metres)
c) 2,324,743,200,000 litres (2.32 billion cubic metres)

- *This is enough urine to fill Loch Ness to a depth of 41 metres.*

And finally, the aforementioned average healthy adult human produces around 200 grams of solid bodily waste a day.

iii. How much is this in a year?

a) 66.00 kilograms
b) 78.50 kilograms
c) 73.05 kilograms

Answers: i) a ii) c iii) c

In extremis, the human bladder can hold 800 millilitres of urine.

An average man is said to weigh 70 kilograms and an average woman 65 kilograms. So every year an average person produces about their body weight in poo.

And some others talk even more ...

How to Build an Environmental Catastrophe

America's kids consume 20 billion disposable nappies between them a year.

Assume children wear nappies for two years.

4.24 million American babies are born annually.

8.48 million American babies are in nappies at any given time.

i. How many nappies is that per baby per day?

a) 7.22
b) 6.46
c) 5.34

And at the beginning it's even more.

How Much Do You Weigh in Flies?

An average housefly weighs about 65 milligrams.

A kilogram of sugar contains 2 million grains.

Answer: i) b

Worldwide, on average, each and every woman will have 2.8 babies.

i. How many grains of sugar does an average housefly weigh?

a) 33
b) 130
c) 307

An average man in the UK weighs 79.2 kilograms.

ii. How many grains of sugar does an average British man weigh?

a) 64,334,000
b) 158,400,000
c) 334,334,000

iii. How many houseflies would it take to balance the scales with an average UK man?

a) 3,003,882.5
b) 1,218,461.5
c) 534,334.8

Cut Throat

Beard hair grows an average of 140 millimetres a year.

Answers: i) b ii) b iii) b

Assuming she was an average 65 kilograms, Lot's wife was turned into 433.3 million grains of salt.

i. If he never shaved, how long would a man's beard grow in a lifetime (say from age 13 to 70)?

a) 26.22 metres
b) 17.31 metres
c) 7.98 metres

A pious Muslim is supposed to sport a beard at least the width of a fist. Assume a man's fist is at least 8 centimetres wide.

ii. How long does it take to grow a Sharia-compliant beard?

a) 184 days
b) 209 days
c) 78 days

A man's beard contains between 7,000 and 15,000 hairs. We'll call it 10,000.

iii. Combining the length of each hair, how much beard does a man grow in a lifetime?

a) 79.8 kilometres (about 50 miles)
b) 67.5 kilometres (about 42 miles)
c) 86.8 kilometres (about 54 miles)

iv. Assuming he shaves, and spends an average of five minutes every day shaving, how long will a man spend shaving in a lifetime?

a) 645 hours (26.9 days)
b) 1,735 hours (72.3 days)
c) 2,813 hours (117.2 days)

Answers: i) c ii) b iii) a iv) b

- *On average hair is 0.2 millimetres in diameter. Men's hair is twice the thickness of women's.*
- *An eyebrow contains about 450 hairs.*

The global wet razor market is worth $6.74 billion every year. Gillette spent ten years and at least $750 million researching its Mach3 razor.

- *Personal earnings of financier George Soros in 2003: $750 million*

Dentures

While George Washington famously wore a set of wooden dentures, before the advent of porcelain false teeth, dentures were often made from teeth pulled from the mouths of the dead. For example, after the battlefield carnage of the American Civil War, teeth extracted from those killed were shipped to England by the barrel load.

Nowadays some 36 million Americans, out of a total population of 295 million, sport a full set of dentures.

There are 6.388 billion people on earth.

i. How many people, pro rata, does this suggest should be wearing a full set of dentures worldwide?

a) 779.6 million
b) 121.5 million
c) 2.1 billion

A full set of dentures consists of 32 teeth.

Answer: i) a

In North Korea the maximum hair length allowed for men under 50 is 2 inches.

ii. So excluding partial dentures, how many individual fake teeth, pro rata, should be in the world's mouths?

a) 3.47 billion
b) 13.34 billion
c) 24.95 billion

(This is equivalent to about four fake teeth for every man, woman and child on earth. Put another way, approximately one tooth in every eight is false.)

Slap Heads in History

About 70 Caucasian men and 35 Caucasian women in a hundred will be affected by baldness to some degree at some point in their lives. Assume this is true for all ethnic groups.

There are 6.388 billion humans. Men outnumber women 106 to 100.

i. How many of the world's men will eventually show signs of baldness?

a) 2.3 billion
b) 4.1 billion
c) 0.8 billion

ii. How many of the world's women will eventually show signs of baldness?

a) 1.09 billion
b) 2.5 billion
c) 0.5 billion

Answers: ii) c i) a ii) a
We each grow twenty 'baby' teeth.

Now assume it was ever thus, and that (as has been estimated) a total of 103 billion people have ever lived.

---

iii. How many bald people have there ever been?

a) 54.6 billion
b) 77.7 billion
c) 32.2 billion

And none of them is more famous than Captain Picard, Patrick Stewart.

The Problem with Women

Is that there aren't enough of them ...

A comparative shortage of younger women is broadly 'normal'. Worldwide, men aged 15-64 outnumber women of a similar age by over 50 million. Perhaps of greater concern, in terms of humanity's future, is that globally there are 50 million more boys than girls aged under 15. The country with the fewest girls per boy is China, where there are twelve boys for every eleven girls – a surfeit of some 17.6 million boys.

Economists seem to be in general agreement that China will become the world's most powerful economy sometime during the 21st century. At this point presumably Chinese men will do what men in the world's wealthier economies have always done to buck this particular demographic trend: they will go looking for their brides overseas.

Here is a partial list of countries and their comparable populations of girls.

Answer: iii) a
A 'full head' is covered by approximately 100,000 hairs.

Country	Number of girls	Country	Number of girls
Spain	2,811,350	France	5,446,716
Nigeria	29,637,684	Romania	1,770,746
USA	29,736,516	UK	5,293,673
Thailand	7,631,337	Italy	3,935,565
Germany	5,879,052	Greece	746,119
Netherlands	1,457,152	Uganda	6,653,764
Burma	5,774,055	Iran	9,411,647

Source: *CIA World Factbook*

Only the USA and Nigeria have populations of young women of sufficient size to match China's 'deficit' alone.

i. What is the minimum number of countries on this list (excluding Nigeria and the USA) with a combined population of young women large enough to meet China's shortfall?

a) 5
b) 3
c) 2

ii. What is the maximum number of countries on this list (excluding Nigeria and the USA) with a combined population of young women large enough to meet China's shortfall?

a) 7
b) 9
c) 11

Answers: i) b ii) a

The world's youngest country is Uganda, where half the population are under 15. By comparison, the world's oldest country is Monaco, where half the population are older than 45. More than 50 per cent of Monegasque women are over 47. Ugandan women have a life expectancy of 46.83 years.

How different Hugh Hefner's life might have been if he'd been born Chinese.

Un Petit Mort

The concentration of human sperm has fallen 29 per cent recently, from nearly 87 million sperm in a millilitre to just over 62 million. Twenty million sperm a millilitre is the lower limit of 'normal'. At orgasm a man produces around 250 million sperm.

i. The *Penguin Atlas of Human Sexual Behavior* claims sexual intercourse takes place, worldwide, 120 million times every day. Assuming an average of one male per coupling and one orgasm per male, how many sperm vie for parenthood every day?

a) 3 billion billion (3 quintillion)
b) 30 million billion (30 quadrillion)
c) 300 trillion

Assume a healthy man can make as many as 1,500 sperm a second.

ii. How many a minute?

a) 150,000
b) 90,000
c) 120,000

Answers: i) b ii) b

Seventy-five per cent of divorced women remarry.

iii. How many a day?

a) 130 million
b) 34 million
c) 175 million

iv. How many a year?

a) 37 billion
b) 47 billion
c) 57 billion

v. There are 2.2 billion adult women on earth. In how many days could one man make them one sperm each?

a) 17
b) 144
c) 203

Frankly it's the kind of noble work to which most men would be proud to lend a hand ...

Meanwhile, a healthy woman with textbook-regular periods menstruates every 28 days (13 times a year) between ages 12 and 49.

How much blood she loses per cycle usually varies from 30–60 millilitres. A heavy period is defined as over 80 millilitres. Assume it's usually 40 millilitres (about the same as a large tot of liquor).

The average number of kids born per woman worldwide is 2.8. So assume an average woman will lose 28 menstruating months to pregnancy during her lifetime.

Answers: iii) a iv) b v) a

vi. How many litres of blood will the average woman menstruate in a lifetime?
a) 18.64   b) 19.68   c) 38.24

(About the same as two cases of wine.)

Twice? A Night?

In 1961 British men married, on average, at the age of 25.6. Nowadays they first marry at an average age of 30.5. And then they die at the age of 75.

i. Assuming the marriage lasts, and such a man could (theoretically) average 'sex' three times a day every day till he dies, how many times will he 'do it' in a lifetime?
a) 48,761   b) 16,608   c) 32,685

According to the *Penguin Atlas of Human Sexual Behavior* the country boasting the longest-lasting sexual intercourse is Brazil at 30 minutes. The US, Canada and the UK follow with 28, 23 and 21 minutes respectively. The quickest sex in the world takes place in Thailand with 10 minutes and Russia with 12 minutes.

Answers: vi) a   i) a

June is the most popular month for Americans to lose their virginity.

ii. At three times a day every day, as above, how many minutes could an average British man spend *in flagrante* in an entire lifetime?

a) 4,246,586
b) 1,023,981
c) 338,442

Put another way that's 1.947 years.

(At a more realistic ten minutes once a month the total number of minutes falls over 99 per cent to 5,340 – just 89 hours.)

Something for the Ladies

Nowadays all email accounts come with a free subscription to adverts offering to increase the size of your willy – which must be a particular challenge if you're a woman.

In 1988 Paul L. Jamison and Paul H. Gebhard published an analysis in the *Journal for Sex Research* of the data collected by Kinsey on penis size.

1 inch is 2.54 centimetres.

	Mean (inches)	Standard deviation (inches)
Flaccid length	3.89	0.73
Flaccid circumference	3.75	0.65
(mean diameter = 1.2 inches)		
Erect length	6.21	0.77
Erect circumference	4.85	0.71
Erectile increase in length	2.30	0.71
Erectile increase in circumference	1.11	0.52

Average erection angle: 15 degrees above horizontal
Average erect diameter: 1.24 inches
Average time to achieve erection: 3 to 8 seconds

Answer: ii) b

Kissing energetically for an hour can burn off over 300 calories.

The data shows that almost 88 per cent of men measure between 5 and 7 inches when erect. When relaxed 90 per cent of men are 3 to 5 inches.

Is this big enough? That depends on two things. First, vagina size. Kinsey reported that the average vagina is 3 inches long with a diameter of 0.8 inches when unaroused, and 3.75–4.1 inches long with a diameter of 2.3–2.5 inches when stimulated (so it's wide enough for birth).

Second, it depends on whether a man's trouser friend works for long enough. Women in Kinsey's studies said that on average they took just under four minutes to achieve orgasm flying solo. Practising with a partner, however, it took between ten and twenty minutes. On average men aged between 16 and 65 can stand this long.

Age group	Median duration of erection in minutes
puberty–15	12.00
16–20	42.88
21–25	54.43
26–30	53.09
31–35	47.24
36–40	40.62
41–45	31.07
46–50	29.02
51–55	21.62
56–60	26.67
61–65	19.50
66–70	7.00
71+	0.00

Source: Kinsey *et al.*, *Sexual Behavior in the Human Male*

> ### i. So (male reader) is your penis long enough?
> ### (Oh all right, female reader too ...)
>
> a) Almost certainly
> b) Yes
> c) Doubtless

Always Thinking About It?

A popular assumption is that men think about sex constantly; there's no hard data, just very often indeed. Complete nonsense of course – otherwise when would they have time to think about football? But for argument's sake let's say it's true, and that they think about it ten times a minute.

Globally, life expectancy at birth, for men, is 65 years. If we assume men start thinking about sex with the advent of puberty, which we'll assume to be at age thirteen, that means they'll think about sex every six seconds for 52 years.

> ### i. If this is true, how many times does a man think about sex in a lifetime?
>
> a) 345.5 million times
> b) 273.5 million times
> c) 654 million times

So quite a lot then ...

But men aren't animals – at least not some of them, after they reach 30 anyway. Let's do the same sum assuming a man only considers sex once every ten minutes.

Answers: i) b i) b

An elephant has a penis 150 centimetres (5 feet) long. A rorqual whale has one a foot thick.

> ### ii. If he thinks about sex every ten minutes, how often will a man think of it in 52 years?
>
> a) 2,734,992
> b) 3,322,543
> c) 6,546,223

There, that's much more civilised.

Will Swap Boyfriend For Chocolate

The deal this tee-shirt slogan posits obviously depends on which boyfriend. If he's a waste of space, a straight swap for a Snickers bar (UK retail price 45 pence, US retail price 80 cents) may satisfy all parties. Alternatively, if he's Good Enough For Now, but probably not The One, then maybe a deal based on his body weight would be better.

> ### i. The average man in the UK now weighs 79.2 kilograms. Assuming we're talking standard-grade chocolate, the kind that retails in a supermarket at £1 ($1.7966) for 200 grams, how much would 79.2 kilograms of chocolate cost?
>
> a) £396 ($711.45)
> b) £764 ($1,372.60)
> c) £1,122 ($2,015.79)

As a solid cube of chocolate, a block of 79.2 kilograms would be 39.35 centimetres (15.5 inches) square. The capacity of the human mouth is about 125 millilitres; 79.2 kilograms is 487.4 bites this size. Eating 217 grams a day (about one big bar), it would take a year to eat.

Answers: ii) a i) a

'Fragrant Flower', the longest python in captivity, is reputedly 14.85 metres (48.7 feet) long.

So, get rid of a decent boyfriend and you'll be pigging out on chocolate for a year.

Things single women want in a man	
Good sense of humour	98%
Intelligence	76%
Good looks	67%
Good at DIY	39%
Able to cook	34%
Well-paid job	32%

Football Crazy

These were the final scores in every World Cup Final since the Second World War.

1950	2-1	1970	4-1	1990	1-0
1954	3-2	1974	2-1	1994	3-2
1958	5-2	1978	3-1	1998	3-0
1962	3-1	1982	3-1	2002	2-0
1966	4-2	1986	3-2		

i. What is the average score in a World Cup Final (approximately)?

a) 3-2
b) 3-1
c) 2-1

Answer: i) b

Per capita consumption of chocolate in the UK is about 9.3 kilograms a year.

These are the final scores in every FA Cup Final since 1950. Replays are marked 'r'.

1950	2-0	1970r	2-1	1988	1-0
1951	2-0	1971	2-1	1989	3-2
1952	1-0	1972	1-0	1990	3-3
1953	4-3	1973	1-0	1990r	1-0
1954	3-2	1974	3-0	1991	2-1
1955	3-1	1975	2-0	1992	2-0
1956	3-1	1976	1-0	1993	1-1
1957	2-1	1977	2-1	1993r	2-1
1958	2-0	1978	1-0	1994	4-0
1959	2-1	1979	3-2	1995	1-0
1960	3-0	1980	1-0	1996	1-0
1961	2-0	1981	1-1	1997	2-0
1962	3-1	1981r	3-2	1998	2-0
1963	3-1	1982	1-1	1999	2-0
1964	3-2	1982r	1-0	2000	1-0
1965	2-1	1983	2-2	2001	2-1
1966	3-2	1983r	4-0	2002	2-0
1967	2-1	1984	2-0	2003	1-0
1968	1-0	1985	1-0	2004	3-0
1969	1-0	1986	3-1		
1970	2-2	1987	3-2		

ii. What is the average score in an FA Cup Final (approximately)?

a) 3-1
b) 3-2
c) 2-1

Answer: ii) c

Since 1980 about twenty amateur soccer players have been killed in the USA by falling goalposts.

Not All Boyfriends Are Equal

Let us now consider the top end of the trade in boyfriends for chocolate; let us assume the boyfriend in question is a bit of a catch. To estimate his true lifetime value we'll have to make some assumptions. First, we'll assume he's 25 years old. Second, that he will avoid redundancy, early retirement, and invalidity. Third, we'll assume he won't have a mid-career crisis, a mid-life crisis, or an affair. Fourth, we'll call him Bob.

Since the legal retirement age in the UK is currently 65, theoretically Bob still has 40 working years left in him, the benefits of which his life partner will share. The average wage for a man in the UK is £28,065 per annum. For a white-collar worker it's £42,900. We'll assume Bob will earn £28,065 pro-rata for half his career, and £42,900 for the rest.

---

i. What is Bob's earning potential at current prices?

a) £854,997 ($1,536,088)
b) £1,419,300 ($2,549,914)
c) £1,633,009 ($2,933,864)

This, we shall therefore assume, is the true lifetime value of a boyfriend.

---

ii. What weight of chocolate in tonnes could you buy with this amount, paying the supermarket price for chocolate of £1 for 200 grams (£5 a kilo)?

a) 134.77 metric tonnes
b) 283.86 metric tonnes
c) 456.22 metric tonnes

Answers: i) b ii) b

Chocolate has a density of 1.3 grams per cubic centimetre. A cube of chocolate with sides of 1 metre would weigh 1,300 kilograms.

> ### iii. How many cubic metres of chocolate would you get from this amount?
>
> a) 382.38 cubic metres
> b) 109.22 cubic metres
> c) 218.35 cubic metres

This corresponds to 1.75 million mouthfuls. Eating a full 125-millilitre mouthful on the hour, every hour, it would still take you over 199 years (about 2.5 lifetimes) to eat.

Alternatively, this many cubic metres of chocolate is equivalent to a cube with sides 6.02 metres (19 feet 9 inches) long - which is about the size of a small house.

Ultimately, then, we can see there's clearly money to be made trading in your old boyfriends. But by Number Freaking we can see that if you get rid of The One - we've called him Bob - you'll be pigging out on chocolate for the rest of your life ...

Boyfriends and Chocolate – Comparative Values

As an economic footnote, finally let us compare and contrast boyfriends and chocolate with other popular assets and commodities.

Answer: iii) c

The global chocolate trade is worth $42.6 billion annually.

i. At the time of writing, the average price of a house in Britain is about £177,474. How many cubic metres of £5 per kilogram chocolate could you buy with £177,474?

a) 27.3 cubic metres
b) 45.2 cubic metres
c) 12.5 cubic metres

This is a cube with sides 3.01 metres (9 feet 10 and a half inches) long – or about the size of a modest bathroom. It's one-eighth the size of the cube of chocolate that Bob our mythical boyfriend is worth.

But is a boyfriend like Bob worth his weight in gold? The price of gold at the time of writing is £226.82 per troy ounce in London and $404.30 in New York. A boyfriend from the UK weighs 79.2 kilograms on average – that's 2,549.61 troy ounces.

ii. What, then, is the average boyfriend worth if weighed in gold (at London and New York prices)?

a) £931,702.21 in London, $1,660,731.80 in New York
b) £578,303.54 in London, $1,030,807.30 in New York
c) £1,617,876.40 in London, $2,883,817.20 in New York

Thus weighed as gold, an average boyfriend is worth over $1 million. By estimating the worth of a boyfriend by his future earnings potential, however, we have discovered that in reality he is worth £1,419,300 ($2,549,914) – nearly two and a half (2.47) times his weight in gold.

Answers: i) a ii) b
One troy ounce is about 31.11 grams.

iii. If a 79.2-kilogram boyfriend is worth £1,419,300, how much is that per kilogram?

a) £26,459.03 ($47,536.29)
b) £17,920.45 ($32,195.88)
c) £9,103.21 ($16,354.83)

- Price of 1 kilo royal beluga caviar: £1,600
- Price of 1 kilo Chanel No. 5 perfume: £9,250
- Price of 1 kilo of cocaine (street prices 2004): London £40,000

Number Freaking conclusion? There's more to be made (potentially) in trading boyfriends than dealing in gold or perfumes.

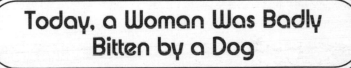

Today, a Woman Was Badly Bitten by a Dog

An internet urban myth: 40 people are sent to hospital because of dog bites every minute. Presumably there are an equal number of male and female victims.

i. How many people receive hospital treatment for dog bites in a year?

a) 14,090,040
b) 21,038,400
c) 37,233,330

Answers: iii) b i) b

In 1992, 5,840 Americans were hospitalised by pillows, and 2,421 by houseplants.

Dog bites man, the old maxim teaches, is not news, but man bites dog is. However there is something interesting about the phrase 'Today, a woman was badly bitten by a dog' (and, indeed, the phrase 'Today a man was badly bitten by a dog') ... In the time it takes to say either of them out loud, if 40 people really are hospitalised by dog bites every minute, it will have happened for real, once.

Worldwide 38 people are bitten by snakes every four minutes. One of them dies.

Go On Do It Again

Britney Spears' first marriage lasted for just 55 hours. Her second has lasted longer.

Britney was born on Wednesday 2 December 1981. The average life expectancy for an American woman is 78 years.

Assume a woman cannot marry before the age of sixteen.

i. How many 55-hour marriages could an American woman like Britney squeeze into a lifetime?

a) 9,881.7
b) 14,256.0
c) 2,114.2

It would be different.

Answer: i) a
A wedding licence in Las Vegas costs $55.

Hard as Nails

Student manicurists are taught that human fingernails grow at a rate of between 0.5 and 1.2 millimetres a week. Assume it's 0.8 millimetres a week.

i. How far is this in a year?

a) 5.67 centimetres
b) 6.74 centimetres
c) 4.16 centimetres

Meanwhile, in another room down the intellectual hallway, student geographers are taught that the earth's surface is made up of giant continental plates which slip and slide past each other. Where they collide, mountain ranges rise like peaks on a rumpled blanket.

Satellite data confirms that the continental plates which support Europe and North America are moving apart at a rate of about 4 centimetres a year. So if you want to see continental drift in action, watch your nails grow.

The earth's circumference is 40,076 kilometres. Another way of saying the continents move apart at 4 centimetres a year is that they are both moving at 2 centimetres per year in opposite directions.

ii. At this rate, how long will it take a particular point on the earth's surface to circumvent the entire globe?

a) 352,440,000 years
b) 12,751,000,000 years
c) 2,003,800,000 years

Answers: i) c ii) c

Tokyo is 1.4 degrees of latitude north of Los Angeles. Stoke is 1.4 degrees north of London.

Eventually, inevitably, America will collide with Japan since they are heading inexorably towards each other at 4 centimetres per year. The distance from Los Angeles to Tokyo is 5,478 miles (8,816 kilometres).

---

iii. How long will it take before Los Angeles hits Japan?

a) 60,300,000 years
b) 495,700,000 years
c) 220,400,000 years

So a little bit after the era after next then. It may not be happening for a while, but when it does? They're gonna be FURIOUS!

Actually California Will Collide with Japan in Just One Era's Time

We may bandy them about willy-nilly in everyday language, but the words we use to describe long periods of time have exact meanings too – just like the words like seconds and minutes that we use for short periods.

- An eon – the longest measure – lasts for billions of years.
- An era spans hundreds of millions of years.
- A period lasts tens of millions of years.
- An epoch lasts millions of years.
- An age spans thousands of years.

(And an age divides into millennia, then centuries, years, months, etc.)

Consequently modern humans exist in a distinct 'chunk' of paleo-archeo-geological time. According to whoever it is that decides this sort of thing, at this time now ...

Answer: iii) c
Birds evolved 100 million years after reptiles: the egg came before the chicken.

We are living in the Third Eon - known, more correctly, as the Phanerozoic Eon.

(The first eon was called the Hadean Eon, the second the Archean Precambrian Eon.)

We are living in the Third Era of the Phanerozoic Eon - also known as the Cenozoic Era.

(The first era of the Phanerozoic Eon was called the Paleozoic Era. The second was called the Mesozoic, and is the era people usually know most about. It started with the Triassic Period - when the continents started to drift, and the first dinosaurs and mammals appeared - before notching up two more periods, the Jurassic and Cretaceous, before the era changed.)

We are living in the Second Period of the Cenozoic Era of the Phanerozoic Eon - also known as the Quaternary Period.

(Like the Tertiary (the first) Period before it, the Quaternary Period has no fancy scientific name. It is a period in history beginning 2 million years ago.)

We are living in the Recent Epoch of the Quaternary Period of the Cenozoic Era of the Phanerozoic Eon.

(The Recent Epoch is everything since 10,000 years ago. Everything before this epoch, the second in this period, was known as the Pleistocene Epoch.)

We are living in the Industrial Age of the Recent Epoch of the Quaternary Period of the Cenozoic Era of the Phanerozoic Eon.

(Other ages that have been documented this epoch are the Old Stone Age, the New Stone Age, the Bronze Age, the Iron Age and the Middle Ages.)

(Alternatively, we are now living in the Latest Age of the Second Epoch of the Second Period of the Third Era of the Third Eon.)

Ussherly Wrong

In 1650 James Ussher, the Archbishop of Armagh, used his bible to estimate the moment of creation as 6 p.m. on Saturday 22 October 4004 BC. By his calculus the universe is just over 6,000 years old.

Acknowledging a flaw in Ussher's estimate, biblical scholars have now recalculated the moment of creation as being more like 11,000 years BP (Before Present, what they call 9000 BC).

Current scientific opinion, however, suggests the earth is a bit older still, and dates it as up to 4.8 billion years old. This means continents have technically had time to travel around the globe nearly two and a half (actually 2.4) times. While this is most definitely *not* what did happen – geologists have shown that the earth's continents coalesced into, then split away from, a single mega-continent called Pangaea – it does provide an interesting perspective on just how old the world is ...

The truth is, any attempt to get a handle on the age of the earth defies human imagination. In part this is because the numbers are mind-boggling, but it's also because so little happened for so long, and humanity and all its endeavours are so relatively recent.

There's a famous analogy which uses an hour on a clock to represent the history of the earth. The problem with this idea is that for the first 10 or 15 minutes absolutely nothing happened. When finally it did, the dinosaurs didn't appear until three or four minutes before the hour, and then died out at a minute to. Our own primate ancestor didn't take its step away from our primate cousins gorillas and chimps until 5 seconds ago, and all of human history has taken place in the last second.

Consider then, instead, this all-new Number Freaking version:

As of Independence Day 2005, at about 5 p.m. local time, America – the Republic – was 7.23 billion seconds old. We'll fix this as our comparative time NOW.

The earth is said to be 4.8 billion years old. So then if we fix 5 p.m. on 4 July 1776 as the relative moment when the earth came into creation, and use this time scale – of 1 year equals about 1.5 seconds – then ...

The oldest rocks would have formed at 6.30 p.m. on 11 May 1819 (the year Queen Victoria was born, Alabama became a state and Spain sold America Florida).

The oldest fossils would date from about 5 a.m. on 2 February 1867 (the year King George V was born, Canada became a Dominion, Nebraska became a state and Russia sold America Alaska – for $7.2 million).

The first dinosaurs would have appeared around 6.30 a.m. on 17 December 1995. The first flowers would have appeared in the late afternoon on 23 January 1999. The dinosaurs would have become extinct just before breakfast on 28 May 2002. Cats evolved on 15 May 2003. Before dogs.

The Grand Canyon began forming just before dawn on 22 March this year (2005). Man's ancestors came down from the trees on the afternoon of 8 April this year. Modern humans first appeared shortly after lunch yesterday. And all known human history began about lunchtime today ...

Merry Christmas

Religious piety is not something of which modern British society could be readily accused. That being said, almost two-thirds of people in Britain are nominally Christian. There are 24.83 million households in the UK, so therefore there are 16.5 million that are nominally Christian.

Now assume all these households buy a Christmas tree. Furthermore, for argument's sake, assume they all buy a real tree as opposed to an artificial one.

Finally, assume a saleable Christmas tree requires a 4-square-metre plot of land to grow in.

> **i. How big a forest would 16.5 million Christmas trees occupy? (We'll forget about the trees doomed to decorate shops and government offices, etc.)**
>
> a) 60 square kilometres
> b) 66 square kilometres
> c) 72 square kilometres

Answer: i) b

- *Area of Guernsey: 63.3 square kilometres*

In the US, 78 per cent of people are nominally Christian and there are 109 million households.

ii. How big a forest would be required to grow a Christmas tree for every nominally Christian household in the US?

a) 340 square kilometres
b) 320 square kilometres
c) 300 square kilometres

- *Area of the Isle of Wight, UK: 381 square kilometres*

Blinded by the Light

Polls in the US indicate that 55 per cent of Americans believe categorically in creationist rather than evolutionary theory. A further 27 per cent believe in evolution guided by divine influence. Only 13 per cent deny creationism absolutely.

There are 240 million Americans aged over 15.

i. If the polls are correct, how many adult Americans believe categorically in Adam and Eve?

a) 214 million
b) 132 million
c) 83 million

- *Total UK population: 60 million*

Answers: ii) a i) b

Every Christmas people in the UK spend a total of £3.9 billion on chilled and frozen food.

These folks also believe that human life began on earth around 10,000 years ago.

And that the first woman was made from a man's rib.

Great Great Great Great Uncle Sam

According to the history books, the Independence of the United States of America was declared 229 years ago late in the afternoon of Thursday 4 July 1776.

This means that in the late afternoon of 4 July 2005, the Republic of the United States of America became:

- 2,748 months old
- 11,949 weeks old (almost)
- 83,640 days old
- 2,007,371 hours old
- 120,442,260 (120.44 million) minutes old
- 7,226,535,600 (7.23 billion) seconds old

Assume that late afternoon means 5 p.m. EDT (Eastern Daylight Time) (9 p.m. GMT).

i. At what time and on what date was the Republic of the United States of America exactly 2 million hours old?

a) 1 a.m. EDT (5 a.m. GMT) Tuesday 31 August 2004
b) 3 p.m. EDT (7 p.m. GMT) Sunday 9 May 2004
c) 10 a.m. EDT (2 p.m. GMT) Thursday 23 September 2004

Answer: i) a

Ten per cent of Americans claim to have seen a ghost.

So happy belated 2 millionth birth-hour for America. (Or happy 120 millionth birth-minute if you prefer.)

America's Birthdays
3,000 months: Saturday 4 July 2026
12,000 weeks: Tuesday 27 June 2006
100,000 days: Tuesday 19 April 2050
8 billion seconds: 11:27:07 EST (Eastern Standard Time) 6 January 2030

The Mother of All Parliaments

Magna Carta was sealed under the boughs of the Ankerwycke Yew at Runnymede 790 years ago on Monday 15 June 1215. The time of day is not recorded, but since we're talking kings and barons and since we can assume there were a fair few hours of debate before hot wax was put to parchment, we shall also assume that like the Declaration of Independence it was sealed in the late afternoon – let's assume at 5 p.m. too – and so began the system of 'modern' English law. So, in the late afternoon of Wednesday 15 June 2005, habeas corpus and trial-by-jury (cornerstones of democratic constitutions throughout the world) will become:

- 9,480 months old
- 41,220 weeks old
- 288,541 days old
- 6,924,992 hours old
- 415,499,510 minutes old
- 24,929,970,000 (24.93 billion) seconds old

America has more gun shops than McDonald's restaurants.

i. At what time and on what date will Magna Carta be exactly 25 billion seconds old?

a) Twenty seconds before 9.27 a.m. Monday 3 September 2007
b) Exactly 2.37 a.m. Saturday 14 June 2008
c) 12.33 p.m. and 38 seconds Sunday 22 October 2006

(Best order the cake now then ... though candles might be a bit of a problem.)

Magna Carta's Birthdays
10,000 months: Thursday 15 October 2048
300,000 days: Wednesday 29 October 2036
7 million hours: 1 a.m. Sunday 5 January 2014

See You There?

The anniversaries of the Declaration of Independence and the sealing of Magna Carta may yet spawn enthusiastic celebrations on their respective sides of the Atlantic. But the question is: will you be here to celebrate too?

The tables on the next page reveal how long British men and women of various ages in 2002 are expected to live. They also show how this is projected to change by 2010.

For example a man who turned 40 in 2002 was expected to live a further 37.44 years (a total of 77.44 years), predicting his year of death as 2039. Thus he'll be around to see Magna Carta turn 300,000 days but not 10,000 months (probably). Likewise, today's teenagers will see the 1000th anniversary of the Battle of Hastings.

Answer: i) a
A century contains 3,100,557,600 (3.1 billion) seconds.

By 2010, however, statisticians predict that the expectation of life for a man turning 40 will have improved to 42.92 years. He should see Magna Carta turn 10,000 months, and the American Republic turn 100,000 days old too.

Of course, nothing is certain. A man of 80 in 2002 is now expected to live until 2009 ... a natural bonus from having reached 80 at all. Put simply, the older you get the longer you'll live. As the numbers in the tables following show, a man born in the UK in 1920 was expected to die by 1976. If he's still alive now he's still expected to live about six years.

Age	Women Life Expectancy		Age	Men Life Expectancy	
	2002	2010est		2002	2010est
0	80.39	~	0	75.68	~
5	75.85	~	5	71.21	~
10	70.89	~	10	66.26	~
15	65.92	~	15	61.31	~
20	61.00	65.76	20	56.47	62.72
25	56.09	60.80	25	51.70	57.77
30	51.18	55.83	30	46.93	52.82
35	46.31	50.87	35	42.17	47.87
40	41.48	45.91	40	37.44	42.92
45	36.72	40.96	45	32.78	37.98
50	32.07	36.06	50	28.62	33.07
55	27.54	31.21	55	23.88	28.22
60	23.18	26.48	60	19.75	23.49
65	19.02	21.94	65	15.93	18.99
70	15.14	17.72	70	12.48	14.87
75	11.67	13.94	75	9.51	11.29
80	8.67	10.71	80	7.09	8.36
85	6.18	8.07	85	5.11	6.09
90	4.31	6.01	90	3.73	4.44
95	3.03	4.48	95	2.63	3.29
100	2.17	3.38	100	1.85	2.52
105	~	2.60	105	~	2.01
110	~	2.07	110	~	1.70
115	~	1.70	115	~	1.52
120	~	0.50	120	~	0.50

A billion seconds is 31 years, 8 months and 8 days. The average UK life expectancy (78 years) is 2.46 billion seconds.

Life expectancy at birth (UK)		
Birth year	Women	Men
1900	50	46
1910	55	52
1920	60	56
1930	63	59
1950	72	66

Life expectancy at birth

Country/Region	Year of birth	
	1998	2003
Western Europe	78	79
North America	76	77
Latin America	69	71
Australia	80	
Asia	65	67
India		63
South Africa		53
Sub-Saharan Africa	49	48

Something Else the French Are Better at Than the British

How long you might live can depend entirely on where you are born. As you peruse the table below, be very glad if you're not a Danish woman. And maybe think about emigrating to France.

Expected Years of Life Remaining at Age 65 by Country (2000)					
	Men	Women		Men	Women
Japan	17.50	22.40	Greece	15.91	18.56
France	17.19	21.63	Norway	15.79	19.68
Switzerland	16.77	20.93	Belgium	15.70	19.65
Australia	16.73	20.23	Austria	15.66	19.61
Sweden	16.65	20.01	Denmark	15.27	17.77
Israel	16.64	18.87	Netherlands	15.13	19.54
New Zealand	16.56	19.93	Finland	15.07	19.18
Italy	16.46	20.57	UK	15.06	18.54
Spain	16.22	20.23	Germany	15.06	18.91
USA	16.02	19.15	Portugal	14.31	18.01
Canada	15.95	19.75	Ireland	14.25	18.05
Singapore	15.92	18.65	Average of all the above	15.91	19.56

Predictions are that over the next 25 years, the structure of world population will shift, with older people an increasingly large proportion of the total. For example, between 1998 and 2025 the world's population of old people aged 65 and over will more than double. Meanwhile the world's population under the age of 15 will grow by just 6 per cent, and the number of children under the age of 5 will grow by less than 5 per cent.

Cash Pyramid

The Great Pyramid of Giza is 137 metres tall. It really is like a small mountain. Built from 2 million blocks of stone, in total the pyramid weighs 6 million tonnes. Most stones weigh 2.5 tonnes, but some weigh up to 50 tonnes. The four right-angled corners align almost exactly with the points of the compass. The sides slope at 52 degrees. The pyramid occupies an area of 45,827 square metres (about 214.1 metres square). It has a volume of about 2.1 million cubic metres, equivalent to a cube with sides of 128.1 metres.

A million dollars in single dollar bills weighs 1 metric tonne. It occupies a space of 1.129 cubic metres.

---

i. How many single dollar bills would it take to construct a pyramid of cash the size of the Great Pyramid at Giza?

a) $767 billion
b) $1.86 trillion
c) $3.83 trillion

Answer: i) b

- *UK GDP: $1.79 trillion*

So, effectively, at current prices the entire British Gross Domestic Product (as single dollar bills) is a pile of cash the size of the Great Pyramid.

Roughly pro-rata ...

- *US GDP ($10.88 trillion): 6 Giza Pyramids*
- *Global GDP ($36.36 trillion): 20 Giza Pyramids*

By the by, it's been estimated that at the rate his fortune is currently growing Bill Gates will become the world's first trillionaire sometime in late 2018. So when might Bill build his own cash Great Pyramid? Probably around Christmas 2020 – 35 years after Microsoft went public.

So Much Cash So Little Time

In fact noted rich man Bill Gates, the Chairman of Microsoft, is so rich he makes Croesus look like an impoverished underachiever. Famous for his philanthropy, the scale of his personal wealth is jaw-dropping. Bill truly has a lot of bills to pay his bills.

According to the Bill Gates Net Worth Page, at the time of writing Bill's personal fortune is worth $31,004,950,033.64 ($31 billion).

A dollar bill costs 4.2 cents to produce, so just to print Bill's fortune in single dollars would cost the US Treasury over $1.3 billion and, at current rates of production, take over seven years.

The Great Pyramid took twenty years to build.

As a pile of single bills Bill's fortune would tower 2,104.19 miles (3,386.36 kilometres) high. That's the height of 8888.1 Empire State Buildings.

Laid end to end these bills would stretch 3,022,129.5 miles (4,835,407.2 kilometres). That's the moon and back 6.45 times.

Stacked in a single block they would fill a cube with sides of 32.72 metres.

They could fill some 240,000 wheelbarrows laid out in a line that would stretch from London to Penzance.

Laid out flat they would cover 123.77 square miles. That's almost the area of the Caribbean island of Grenada.

At current prices the US invasion of Grenada cost $142.59 million. So not only could Bill afford to invade Grenada 217 times, but if he felt more in the mood for a bit of landscape design on invasion morning, he could paper over almost the entire island with dollar bills instead.

The IKEA store at Thurrock in Essex, UK, (for example) has a floor area of about 23,500 square metres. Bill's fortune could fill it to a depth of 1.49 metres (the height of a twelve-year-old boy).

Since mid-March 1986, Bill has been accumulating this wealth at an average rate of $55.05 per SECOND.

i. If Bill decided to watch the three movies in Peter Jackson's $297 million *Lord of the Rings* trilogy back to back (total running time 559 minutes), how much would he earn while viewing?

a) $166,023
b) $1,846,377
c) $4,891,505

Answer: i) b

Bill's wealth grows so fast he could have financed all three movies in about two months.

How Much Money Has There Ever Been?

The total of all the wealth in the world is said to be $51.48 trillion. This includes both money we might regard as our collective savings and money we might regard as our collective earnings. The savings part amounts to about $15.12 trillion. The earnings part, the current global GDP, is $36.36 trillion. GDP is a measure of all the global wealth produced in a given year. In 1950 it was about $5.33 trillion.

The average of $5.33 trillion and $36.36 trillion is about $20.85 trillion. Assume it is a yearly average global GDP for an average global lifetime of 67 years.

i. What is the total GDP for the current global population for an entire lifetime?

a) $390 trillion
b) $802 trillion
c) $1,396.95 trillion

To put a value on how much money there might have ever been – or at least a maximum estimate – multiply this number by how many people have ever lived, then divide it by the current global population.

The total number of humans who've ever lived is estimated to be between 96 and 110 billion, including the 6.388 billion currently alive. Assume the total is 103 billion.

Answer: i) c

US dollar bills are 2.61 inches wide, 6.14 inches long and 0.0043 inches thick.

ii. So how much money has there ever been?

a) $12,337,342,000,000,000,000 ($12.3 quintillion or $12.3 million trillion)
b) $32,889,002,000,000,000,000,000,000 ($32.9 septillion or $32.9 trillion trillion)
c) $22,524,397,000,000,000,000 ($22.5 quadrillion or $22.5 thousand trillion)

The sheer volume of this much money, say in single dollar bills, would be staggering. A million dollars in single bills has a volume of 1.129 cubic metres. All the money there's ever been would have a volume of over 25 cubic kilometres. And how much is this? Well, on average water pours over Niagara Falls at 6,000 cubic metres per second. That's enough to fill four Olympic-sized swimming pools 60 times every minute, day in and day out. All the money there has ever been would occupy the same volume as the total amount of water that crosses Niagara Falls in seven weeks.

But obviously, for most of history, annual global GDP has been less than $5.33 trillion. So if this Number Freak feels like an overestimate, consider the following: just as the scale of GDP has exploded so has global population and these two counterbalance each other in the equation. Furthermore, as we all know, money isn't worth what it was: in 1950 $5 trillion was worth about the same as $39 trillion is now. So perhaps a $20.85 trillion average might even be an underestimate ...

- In 1750 per capita GDP in Britain, in 2004 money, was about $2,170 (about the same as modern Pakistan).
- In 1750 per capita GDP in France, in 2004 money, was about $1,500 (about the same as modern Kyrgyzstan or Haiti).
- All the money there has ever been (in dollar bills) could build 12,110 Giza Pyramids.

And as it says in the song: if you divided it equally between the world's richest men there still wouldn't be enough to go round ...

Answer: ii) c

The search term 'How much money has there ever been?' yields zero results on Google.

It's Pronounced Lar-Tay

Four out of five people now visit a coffee shop at least once a week, with the serious addicts – one in five – visiting every day.

In the UK there are 35.5 million adults aged 18 to 64.

i. Between them how many visits do they make to coffee shops every year?

a) 456 million
b) 3.701 billion
c) 6.445 billion

Around half of these customers spend at least 30 minutes over their coffee. The rest? Assume they must average at least five minutes. Assume they all order a coffee.

ii. Therefore how long do all these people combined spend over coffee every year?

a) 3,567,881,004 hours
b) 1,079,458,333 hours
c) 2,482,565,221 hours

This is the equivalent of about 478,500 average UK working years of 2,256 hours. The average annual salary is £25,170.

iii. How much might coffee drinkers earn if they weren't slurping?

a) £12,043,845,000 (£12 billion)
b) £24,334,878,000 (£24 billion)
c) £42,228,673,000 (£42 billion)

Answers: i) b ii) b iii) a

About half of all coffees sold are lattes – that's 1.856 billion cups. A cup is 284 millilitres.

iv. How many litres of lattes do these bean heads consume every year?

a) 145.6 million litres
b) 329.5 million litres
c) 525.7 million litres

This would fill about 350 Olympic-sized swimming pools.

More Tea Vicar?

According to the BBC, between them, people in the UK consume 196 million cups of tea daily. Ninety-eight per cent are drunk with milk.

i. How many cups of tea do people in the UK drink every year?

a) 242,445,000,000 (242.4 billion)
b) 71,589,000,000 (71.6 billion)
c) 316,016,000,000 (316 billion)

On average a cup of tea contains around 45 milligrams of caffeine.

ii. So how many metric tonnes of caffeine do British people between them consume annually from tea?

a) 1,445.6
b) 8,855.2
c) 3,221.5

Answers: iv) c i) b ii) c

Americans drink an average of 1.52 cans of 'soda' a day each.

Which weight for weight is about four times the global production of cocaine.

> iii. In UK cooking parlance, one cup is 284.1 millilitres (half a UK pint).
> On this basis, how many litres of tea are consumed in the
> UK every year?
>
> a) 78,339,878,000 litres
> b) 44,782,556,000 litres
> c) 20,338,434,000 litres

An Olympic-sized swimming pool holds 1.5 million litres.

> iv. How many Olympic-sized swimming pools of tea do people in the
> UK collectively drink every year?
>
> a) 5,345
> b) 13,559
> c) 28,654

According to the Amateur Swimming Association there are 1,400 public and 1,600 school swimming pools in England. This suggests a pro-rata total of about 3,600 publicly accessible pools in the UK. Of course this takes no account of private pools in private homes or water parks, but equally most pools are not Olympic-sized. So let us assume there are the equivalent of 4,000 (so as not to underestimate) Olympic-sized pools in the UK.

> v. How many times could the UK's annual tea consumption fill the
> nation's swimming pools?
>
> a) 0.45
> b) 2.04
> c) 3.39

Answers: iii) c iv) b v) c

(Add two sugars per cup? That's 143.2 billion lumps or 715,890 metric tonnes of sugar.)

Sniff

Global production of export grade cocaine in 2003 was estimated to be 800 metric tonnes (a fall from 827 metric tonnes the year before).

The street price of cocaine is about $71,500 (£40,000) per kilo.

---

i. What would the total retail value of the global cocaine trade appear to be?

a) $65.3 billion
b) $68.2 billion
c) $57.2 billion

- *GDP Romania 2003: $60.4 billion*

But recent figures from the United States Office of National Drug Control Policy valued the 260 metric tonnes of cocaine smuggled into the US at $35 billion, which would value 800 tonnes at $107.7 billion. This is $134,625 (£75,000) per kilogram.

- *GDP Malaysia 2003: $103.2 billion*

There are several explanations which might combine to account for this discrepancy. One is that cocaine dealers are murderous thieving scum who swindle their customers. Another is that cocaine is routinely adulterated with other substances – for example caster sugar.

Answer: i) c

The Irish drink 22 per cent more tea per head than the British.

ii. So just working with the latter, what percentage of the cocaine sold on the streets is, on average, actually cocaine?

a) 53.1 per cent
b) 75.2 per cent
c) 33.9 per cent

Cocaine, they say, is God's way of saying you've got too much money ...

iii. Because what price, approximately, is a cocaine user paying, at street prices, pro-rata, for the caster sugar they buy mixed with their cocaine?

a) $33,500 per kilogram
b) $71,500 per kilogram
c) $112,600 per kilogram

- *Caster sugar – approximate supermarket retail price per kilogram: $1.45*

This is more than a 88,000-fold mark-up.

iv. Two teaspoons of sugar weigh about 10 grams. At the price it costs mixed with cocaine, about how much would two teaspoonfuls of sugar in a cup of coffee cost?

a) $715
b) $1,132
c) $234

A 500-gram bag of sugar is said to contain 1 million grains. So 10 grams must contain 20,000 grains. So how much would the sugar in your coffee be at the price it is when it's mixed with cocaine? About 2 UK pence or 3.6 US cents - a grain.

Answers: ii) a iii) b iv) a

Two per cent of UK adults admit to using cocaine recently.

Drugs 'n' Money

On the New York market, at the time of writing, the price of a kilo of gold (without discounts for a bulk purchase) is $13,015.24. Which makes pure cocaine, at up to $135,000 a kilogram, very expensive indeed.

But actually, as drugs go, cocaine's as cheap as chips.

Pamidronic acid is the key ingredient of an important anti-cancer drug. To make this drug, pamidronic acid is mixed with just one other chemical, which can be bought on the world market for less than $1 a kilogram (at current prices). The market for this drug is valued at $445 million per year. The entire annual global supply of the drug is produced from just 66 kilograms of pamidronic acid.

---

i. By the time it is converted into the drug, what is the pro-rata value of a kilogram of pamidronic acid?

a) $13.48 million (100 times the price of cocaine)
b) $6.74 million (50 times the price of cocaine)
c) $1.35 million (10 times the price of cocaine)

- Bulk price of ingredients to make one kilogram of pamidronic acid: $50
- Multiple of basic ingredients cost for finished product: 134,800

Fluconazole is a drug used to treat foot fungus. It can also be used to treat several secondary infections of Aids. A 150-milligram capsule of the world's leading fluconazole brand can cost up to $27. There are 1,000 milligrams in one gram.

Answer: i) b

ii. Bought as 150-milligram capsules, how much would a kilogram of the world's leading brand of fluconazole cost?

a) $180,000 (one and a third times the price of cocaine)
b) $139,000 (just over the price of cocaine)
c) $424,000 (over three times the price of cocaine)

- *Minimum price per kilogram of fluconazole on the world market: $100*

So what can be more lucrative than the trade in illegal drugs? The trade in legal drugs.

But if you really want to make money – become a diamond dealer.

- *Value of a kilogram of cut and polished one-carat diamonds: $80 million*

Does Crime Pay?

Numbers are, of course, hard to come by. Witness estimates of fraud in the UK, which range from as little as £5 billion to as much as £16 billion per annum. The UK drugs trade, meanwhile, is estimated to employ, at least occasionally, upwards of 20,000 people in London and 100,000 throughout the UK. It is valued at about £8.5 billion, annually, at current prices.

The Assets Recovery Agency (ARA), the law enforcement agency in the UK charged with recovering ill-gotten gains, are reportedly targeting 400 career criminals with assets of at least £440 million ($791 million) between them.

Answer: ii) a
One kilo of Kopi Luwak, the world's most expensive coffee, costs £560.

> ### i. What is the value, on average, of the alleged assets of the 400 criminals targeted by the ARA?
>
> a) £3.8 million ($6.83 million)
> b) £1.1 million ($1.98 million)
> c) £267,000 ($480,000)

The average annual wage in the UK is £25,170. Assume a full UK working life is 47 years (from age 18 to 65). Assume that at current prices average gross lifetime income is a multiple of the two.

> ### ii. What is the average gross lifetime income in the UK?
>
> a) £1.183 million ($2.125 million)
> b) £3.432 million ($6.166 million)
> c) £2.454 million ($4.409 million)

Therefore on average the career criminals targeted by the ARA reported here have assets equal to 93 per cent of all an average person earns in a lifetime before tax is deducted.

So what is the answer to the question does crime pay? The answer is not bad at all ...

Answers: i) b ii) a

An average FBI fugitive will travel 969 miles to escape but will be caught after 157 days (five months).

Whoosh

Hurricanes are categorised by their wind speeds.

Hurricane strength	Associated wind speed (mph)
Level 1	74–95
Level 2	96–110
Level 3	111–130
Level 4	131–155
Level 5	156+

Nervous impulses travel through our nerves at about 70 metres per second.

i. What speed is this in miles per hour?

a) 94.6 mph
b) 156.6 mph
c) 138 mph

- *Wind speed of a sneeze: 93.2 mph*

Zap

There are estimated to be 40–60 flashes of lightning (somewhere in the world) every second. Call it 50.

Answer: i) b

The Atlantic hurricane season officially starts on 1 June.

i. How many lightning flashes are there every year?

a) 576 million
b) 1.58 billion
c) 1.78 billion

A flash of lightning happens when positive electrical charge on the ground discharges upwards into a thundercloud. The flash has a diameter of about 1 inch. The temperature at the point of discharge is up to 28,000°C, so any insects, worms or small mammals in the immediate vicinity are literally toast. A typical flash appears to last about a fifth of a second, but may be made up of over 200 faster flashes too rapid for the eye to see. Thunder is the air rushing in to fill the vacuum it creates. The flash has a search radius of about 50 feet. If you're the largest object within that radius, you're toast too.

Some 30 per cent of people struck by lightning die, very often of heart failure, and mostly within an hour. Three-quarters of survivors are permanently disabled. Those most at risk are young men goofing around outdoors on summer evenings. Excluding those killed indirectly, by a tree felled by the bolt, say, on average about three people in the UK and 80 people in the USA are killed by lightning every year.

So there you have it, young women are at mortal risk from childbirth and young men from playing football in the rain.

Why You Should Always Take an Umbrella

At any given moment there are estimated to be 2,000 thunderstorms worldwide. How long a thunderstorm lasts depends on various things –

Answer: i) b

A typical lightning bolt generates a current of 5,000–20,000 amps.

including where it is. A study in Eastern Europe calculated the average thunderstorm to last for about two and a quarter hours. Another, in the US, calculated it to be three and a half to four hours. For the purpose of Number Freaking we will assume the global average is three hours.

---

i. How many thunderstorms occur worldwide every year?

a) 5,844,000
b) 14,657,000
c) 23,342,000

A single thunderstorm can release over 450 million litres of water.

---

ii. Globally, about how many litres of rain are unleashed annually by thunderstorms?

a) 2,630 trillion litres
b) 5,245 trillion litres
c) 9,332 trillion litres

- *Capacity of the English Channel: 4,855 trillion litres*
- *Capacity of Lake Michigan: 4,814 trillion litres*
- *Capacity of Lake Ontario: 1,541 trillion litres*

Thunderstorms are not the same as simple rainstorms. They do not account for all the rainfall that falls on earth. They can occur anywhere on land or sea, but are obviously less frequent in, say, the Sahara than the Mid-West. The surface area of the earth is 510,065,600 square kilometres.

Answers: i) a ii) a

Energy from a major thunderstorm could supply all the USA with electricity for twenty minutes.

> **iii.** If all the rain precipitated annually by thunderstorms worldwide fell equally across the earth's surface how deep would the water be?
>
> a) 12.67 millimetres
> b) 24.33 millimetres
> c) 5.16 millimetres

Why should you always take an umbrella? Because it might rain.

Blah Blah Blah Blah Blah

The latest rather charming theory of how humans came to speak, but other apes did not, argues that after we stood up and (critically) lost our hair, our babies, which once could cling on, now had to be carried, obliging parents to put down their offspring while foraging and reassure them vocally.

A broadcast news rule of thumb is that a news anchor reads written text aloud at the fast but intelligible pace of three words per second. Everyday conversation could be the same, but often isn't. So assume human speech averages out at two words a second.

So how much time per day do we spend talking? Apparently two-thirds of us now prefer chit-chat to substance. Language may be the glue of human bonding, but if on some days we can feel positively chirpy, on others we could barely mumble 'Help!' if we were falling down a hole ...

A survey somewhere once claimed the average married couple talk for only ten minutes per day. That's just five minutes each if equally shared ... But a day presents many conversational opportunities. Whether it's shouting at our kids, collecting the dry cleaning or remonstrating with another drunk in a bar, we'll assume that on average we each spend 60 minutes with words coming out of our mouths every day.

Answer: iii) c

i. So how many words does a person speak in a day?

a) 13,300
b) 7,200
c) 10,100

ii. How many is that in a year?

a) 2,629,800
b) 3,554,200
c) 5,685,300

iii. And assuming a talking life of 73 years – in the UK at least – how many words will you speak in a lifetime?

a) 191,975,400
b) 232,866,300
c) 365,200,300

The richest person on earth would be the one with a patent on the word 'the'.

Statistically, the fifteen most frequently used words in a language will account for 25 per cent of text. The 100 most frequently used will account for 60 per cent, the thousand most frequent for 85 per cent, and the 4,000 most frequent for 97.5 per cent.

The fifteen most frequently written words in English are:

1. the	4. in	7. for	10. that	13. he
2. of	5. and	8. was	11. on	14. with
3. to	6. a	9. is	12. at	15. by

The most frequently used letter in English is 'e', followed by 't'.

Answers: i) b ii) a iii) a

The Hawaiian alphabet has twelve letters.

Read Read Read Read Read

A novel is normally assumed to be a fictional manuscript in excess of 50,000 words. In the genre of factual books, a publisher might reasonably ask for between 80,000 and 100,000 words. A children's book can contain less than 50 words. This book contains about 43,000. Likewise books may not be just words. A picture can paint a thousand words (apparently) and a decent graph even more. Like a conversation there's no fixed length for a book. But let us assume an average book is all words and has 350 words to a page and 200 pages, a total of 70,000 words.

In 1999, when somebody bothered to count, it was estimated that the total number of volumes physically shelved in British public libraries (i.e. actual bound books, not titles) was 128 million – a fall of 9 per cent on the previous decade.

i. Assuming it's still 128 million, how many books are held in British libraries for each of the 55 million Britons of reading age (which we'll assume is five plus).

a) 9.44
b) 4.02
c) 2.33

ii. At 70,000 words a book, how many words are printed in the 128 million volumes stocked in British public libraries?

a) 8,960,000,000,000 (8.96 trillion)
b) 16,334,000,000,000 (16.33 trillion)
c) 24,778,000,000,000 (24.78 trillion)

Answers: i) c ii) a

iii. If the entire UK reading population spent the average 16.9 hours they're awake reading aloud, how long would it take them (at two words a second) to read the entire stock of British public libraries?

a) 1 day, 5 hours, and 44 minutes
b) 2 days, 14 hours and 22 minutes
c) 4 days, 20 hours and 37 minutes

Which book would you start reading first?

The world's twelve most commonly spoken languages

1. Chinese	5. Arabic	9. Japanese
2. Hindi	6. Bengali	10. German
3. English	7. Russian	11. French
4. Spanish	8. Portuguese	12. Punjabi

This Too Shall Pass

(a phrase said to be a summary of all human wisdom)

Rightfully renowned for its scholarly authority, at the time of writing, the *Encyclopaedia Britannica* consists of 32 volumes containing 33,114 tissue-thin pages, on which are printed about 44 million words in around 65,000 articles. The online version has around 75,000 entries.

By comparison, the online encyclopaedia *Wikipedia*, still proving its credentials, contains some 250 million words in 360,000 entries, and continues to grow by a continuous process of updating, refinement and revision.

Answer: iii) a

The Piraha Amazonian Indians have no words for numbers bigger than two.

Suffice to say, both encyclopaedias are full of a lot of stuff.

i. How long would it take to read *Wikipedia* aloud (continuously at two words per second)?

a) 12.5 years
b) 4.0 years
c) 3.4 years

ii. How long would it take to read *Encyclopaedia Britannica* aloud (continuously at two words per second?)

a) 0.7 years
b) 4.8 years
c) 6.2 years

Only two people (other than the editor, one presumes) are known to have read the whole of the *Encyclopaedia Britannica*. One's an American journalist, who took a year – and, it may not surprise you to learn, wrote a book about it. The other was a 75-year-old pensioner named Ron Heap from Thurnby in Leicestershire, UK, who took ten years. Well done Ron.

How Much Does Television Cost?

Of the BBC's annual spend on programmes of £2.36 billion ($4.24 billion), £812.3 million went to BBC1 TV. Including news programmes BBC1 broadcasts 24 hours a day all year round.

Answers: i) b ii) a

In spoken Japanese, 19 sounds like the Japanese for bad luck, 33 like trouble, and 42 like death.

i. How much does BBC1 spend on programmes on average per hour?

a) £265,445.76 ($476,899.85)
b) £134,772.88 ($242,132.95)
c) £92,664.80 ($166,481.57)

At the time of writing the average house in the UK costs £177,474 to buy.

ii. How many hours of BBC1 programmes would it take to buy a house?

a) 3.453
b) 1.915
c) 0.557

This is about the original output in hours of a popular BBC1 TV soap opera like *Eastenders* every week. So, build a house or make an *Eastenders*... You decide.

The One Where They Swept Us All Off Our Feet

Friends was one of those global television hits where everyone had a favourite character. Which was yours?

The US audience for the final episode was recorded at 52.5 million; that's somewhere between the populations of South Korea and Italy. The surprising thing, in a way, was how all those people fell in love with Monica, Chandler, Joey, Rachel, Ross and Phoebe in so little time.

Answers: i) c ii) b

Four thousand Americans are injured by tea pots every year.

There were a grand total of 238 episodes. Assume each episode was about 25 minutes long (after all the ad breaks were removed).

i. What would that make the total running time of *Friends*?

a) 71 hours 15 minutes
b) 99 hours 10 minutes
c) 103 hours 20 minutes

There are 16.9 waking hours in an average waking day.

ii. So for how many waking days did we actually know these comic chumsters in total?

a) 13.35
b) 5.87
c) 9.13

So barely a couple of long weekends really, and yet so many miss them so much ... The words 'life', 'a' and 'get' spring to mind.

By the same token, at the time of writing there have only ever been 335 episodes of *The Simpsons*. Assume they have all been 25 minutes long too.

iii. What has been the total running time of *The Simpsons*?

a) 139 hours 35 minutes
b) 117 hours 5 minutes
c) 84 hours 10 minutes

Answers: i) b ii) b iii) a

On average there are five TV sets for every six people in the US.

Assume again 16.9 waking hours in an average waking day.

> **iv. For how many waking days have we all known the Simpsons?**
> **(And yet we feel we know them so well.)**
>
> a) 20.22
> b) 14.91
> c) 8.26

Only about a week? Marge?!

Bladerunner

Some say we'll all be walking into the future hand-in-pincer with advanced autonomous intelligent robots.

Not all of us we won't ...

To date 720,000 working robots are operating worldwide, 57 per cent of which (410,000) are to be found in Japan, improving the productivity of its aging work force, and enhancing the care of its elderly.

The population of Japan is 127 million.

> **i. How many people is that per robot in Japan?**
>
> a) 1,453
> b) 310
> c) 28

ii. How many robots would there need to be in our world of 6.388 billion for there to be one 'working robot' for every 310 people?

a) 20.6 million
b) 43.5 million
c) 65.1 million

They won't all be built by the end of next week then ...

Only Connect

By the end of the 1990s people in the developed world made more phone calls each year than were made throughout the entire 1980s.

Global Telephones	World	EU	US	Japan	China	'Majority'
Population (millions)	6,338	456	300	127	1,299	4,206
Landline phones (millions)	844	239	182	71	263	89
Mobile phones (millions)	1,095	315	159	87	269	265
Total phones (millions)	1,939	553	340	158	532	356

300 million Americans share 182 million landline phones.

i. How many Americans are there per landline phone?

a) 1.65
b) 4.65
c) 2.43

Answers: ii) a i) a

In computer jargon 'a jiffy' is 10 milliseconds.

4.206 billion people – nearly two-thirds of the world's population – do not live in the EU, the USA, China or Japan. Those who live in South America, Africa, Eastern Europe and India, for example. We'll call them 'the majority'. They share 89 million landline phones.

ii. How many of the majority are there per landline phone?

a) 47.26
b) 123.56
c) 76.33

456 million Europeans share 315 million mobile phones.

iii. How many Europeans are there per mobile phone?

a) 10.89
b) 4.77
c) 1.45

The 4.206 billion majority share 265 million mobile phones.

iv. How many of the majority are there per mobile phone?

a) 15.87
b) 22.45
c) 38.22

- *Two-thirds of the world's population have access to only one-fifth of the world's phones.*
- *Averaged worldwide there are enough phones for every 3.29 people to share one.*

Answers: ii) a iii) c iv) a

In 57 per cent of American households there are three or more phone handsets.

falling from the World's Tallest Building Could You Phone a friend?

Suddenly, you find yourself falling from the world's tallest building – the Taipei 101 building. It is 508 metres tall.

Terminal velocity for a human with arms and legs outstretched is about 200 kph or 55.6 metres per second. If we assume that you accelerate to this speed at 9.8 metres per second per second (which is the speed gravity pulls objects towards the earth) and then fall at this constant rate, it takes a fall of 202.6 metres and a shade under 5.95 seconds to reach terminal velocity. Assume you don't hit the sides on the way down.

i. How long will it take you to hit the ground?

a) 11.5 seconds
b) 6.4 seconds
c) 23.6 seconds

Will you have time to ring a loved one as you fall? Tell them where you are and what you're doing? Tell them you love them? Assume you hit the speed dial as you jump and they await your call, phone in hand.

ii. Will you have time to say, specifically, 'I'm falling from the world's tallest building. I love you. Goodbye' (assuming you speak at three words per second)?

a) Yes
b) No
c) Could be yes, could be no.

Answers: i) a ii) a

(You could test this one out by phoning your loved one from the ground, saying these things, and timing yourself. You may wish to phone them back straight afterwards to reassure them.)

Warning: you probably wouldn't have time to send a text message.

Time it would take to hit the ground if you jumped from ...

- *Giant Sequoia (Redwood) (94.5 metres): 3.92 seconds*
- *Big Ben (96.3 metres): 3.96 seconds*
- *George Washington (on the scale of the Mount Rushmore face) (137.2 metres): 4.80 seconds*
- *Blackpool Tower (158.1 metres): 5.19 seconds*
- *Eiffel Tower (300 metres): 7.70 seconds*
- *Empire State Building (381 metres): 9.16 seconds*
- *South Deep Gold Mine (in South Africa, the world's deepest single drop mineshaft) (2,992 metres): 56.12 seconds*
- *The height of Mount Everest (8,848 metres): 2 minutes 41.44 seconds*
- *The depth of the Mariana Trench (10,924 metres): 3 minutes 18.78 seconds*

A.S.

The highest point on the earth's surface is Mount Everest. The deepest is the Mariana Trench in the Pacific. The difference in altitude between the two is 19,772 metres. A human dropping this distance in freefall at the human terminal velocity of 200 kph would take a few seconds under six minutes to reach bottom.

In the pantheon of cartoon comedy, the notion that one might burrow through the earth from, say, London and come out on the other side in Australia is a classic ...

On an average working day in New York City people make over 36 million phone calls.

The equatorial diameter of the earth is 12,756.28 kilometres. The polar diameter is 12,713.51 kilometres. The average of the two is 12,734.9 kilometres. Imagine, then, that a tunnel has opened up through the earth 12,734.9 kilometres long, and, oops, would you believe it, you've only gone and fallen down it ...

---

i. Falling at terminal velocity how long would it take, theoretically, for you to fall through the earth?

a) 79 minutes
b) 17.5 hours
c) 2 days 15 hours 40 minutes 29 seconds

Virtually long enough to earn a salary ...

Tommy This and Tommy That

Some of the numbers (and assumptions) in this book might be very wrong indeed. But if you find one, don't bother writing in. Number Freaking isn't for reference, it's for entertainment. It's an attempt to put numbers where numbers don't exist – or where it might just be fun if they did. But sometimes, in its absurd, perverse and crazy way, the process of Number Freaking reveals a truth about the world we didn't know before ...

Defence of the realm is the first duty of any government. Currently the UK maintains a professional army of 111,780 soldiers consisting of 14,500 officers (including 1,480 women) and 97,260 'other ranks' (including 6,760 women). One in eight soldiers is therefore a boss.

So then, let us imagine the enemy is at the gate ...

Answer: i) c

It would take sound waves 3.5 hours to travel from New York City to San Francisco.

i. The coastline of the UK is 12,429 kilometres long. If the entire current army were deployed to patrol the country's shores, how far apart would each soldier be spaced?

a) 111.2 metres
b) 202.5 metres
c) 38 metres

Which is fine and dandy if the attack comes by sea, but what if it comes by air?

ii. The UK has an area of 244,030 square kilometres. Deployed equally distant, how big an area would each soldier need to defend?

a) 24.63 square kilometres (about 4.96 kilometres square)
b) 2.18 square kilometres (about 1.48 kilometres square)
c) 13.63 square kilometres (about 3.69 kilometres square)

- *Size of an average IKEA store: 0.0214 square kilometres*

Proof by Number Freaking: British people can all sleep safe in their beds.

(Actually, some can sleep safer than others. There are 0.0018736 soldiers per head of the UK population. But David and Victoria Beckham, for example, have two bodyguards each. Which means they're 1,067 times better protected than the rest of us.)

Tudor Wireless

The distance of the horizon (a critical factor for a sentry) obeys a simple formula: the distance to the horizon in miles is the square root of the height of the observer's eyes (or the top of an object) in feet, multiplied by 1.22.

Answers: i) a   ii) b

In the 1990s Britons claimed 30 UFO sightings a week. Since 9/11 reports have been almost zero.

i. Standing by the water's edge how far could a 5 foot 9 inch sentry see out to sea? (His eyes we'll locate at 5 feet 6 inches high.)

a) 8.66 miles (15,242 yards)
b) 4.84 miles (8,518 yards)
c) 2.86 miles (5,034 yards)

And if he stood on a hill he'd see even further.

When Elizabeth I had a line of beacons built across Britain to warn of any approaching armada that might happen by, let us assume they were 30 feet high and stood on hills at least 100 feet high.

ii. At what distance would the top of such a beacon be visible on the horizon to a 5 foot 9 inch observer?

a) 48.9 miles
b) 32.6 miles
c) 16.8 miles

Nowhere on the British mainland is much more than 540 miles from London, so it would take a line of no more than 33 beacons to send a warning to London. If we assume such beacons could be lit within two minutes, it took no more than 66 minutes for the message that the Spanish were coming to arrive. Overall speed? 491 mph.

Answers: i) c ii) c

At 10,000 feet a pilot can theoretically see 122 miles in every direction (46,778 square miles).

Brothers in Arms I

The USA has an 'edge' of 31,958 kilometres – a coastline of 19,924 kilometres plus 12,034 kilometres of borders – and an army of 1.2 million, made up of around half a million soldiers on active service, backed up by 700,000 National Guards and reservists.

i. As with the UK, if the entire army were deployed to patrol the country's rim, how far apart would each soldier be spaced?

a) 75.09 metres
b) 26.63 metres
c) 117.22 metres

ii. The USA has an area of 9,363,123 square kilometres. Deployed equally distant, how much land could each soldier guard?

a) 35.65 square kilometres (5.97 kilometres square)
b) 18.52 square kilometres (4.30 kilometres square)
c) 7.80 square kilometres (2.79 kilometres square)

And so (writes a retired Brigadier from Tunbridge Wells), Americans are four times better protected than Britons if invasion comes by sea, but Britons are 3.5 times better defended than Americans if it comes by air.

Answers: i) b ii) c

In 1789 the US Army numbered just 840 men.

Brothers in Arms II

The size of the entire Australian National Defence Force (army, navy and air force) is 53,000 servicepersons.

> **i. The Australian coastline is 25,760 kilometres long. So how much coastline is that per Australian military person?**
>
> a) 486 metres
> b) 1,512 metres
> c) 888 metres

> **ii. The area of Australia is 7,617,930 square kilometres. So how big an area is that per Aussie Anzac?**
>
> a) 332.78 square kilometres (18.2 kilometres square)
> b) 484.44 square kilometres (22 kilometres square)
> c) 143.73 square kilometres (12 kilometres square)

While Australia lacks the military personnel to scan its entire land area the problem could be solved if all those under orders were to stand on a chair on top of a table.

Until then? Australia might be at serious risk of attack ...

Answers: i) a ii) c

From Kosciusko, Australia's tallest mountain (at 7,310 feet), one can technically see 104.3 miles.

Too Late, Australia Has Already Been Invaded

In August 2004 residents of Melbourne learnt they'd been invaded by an ant colony spanning 100 kilometres (62 miles).

Brown Argentine ants, the ants in question, are about 4 millimetres long and weigh 2.5 milligrams. In their native Argentina, colonies compete violently, but outside Argentina they co-operate with their kin instead, forming what are known as super colonies or, more contentiously, super organisms.

The biggest of these super colonies ever found was discovered in 2002. Entomologists found a network of co-operative Argentine ant colonies, consisting of billions of animals in millions of nests, spanning 3,700 miles from the north of Italy to the Atlantic coast of Spain. In Japan, meanwhile, a super colony of Japanese red wood ants was found to span 12.4 miles, and estimated to contain 45,000 nests, and 307 million animals (306 million workers and 1.1 million queens).

At 62 miles across, this suggests that the Melbourne colony is 25 times the size of the Japanese colony.

i. If the Japanese colony contains 307 million animals, how many ants does it suggest have invaded Melbourne?

a) 21.028 trillion
b) 7.675 billion
c) 63.874 billion

Answer: i) b

Ants are thought to have first evolved on earth 100 million years ago.

ii. If each ant weighs 2.5 milligrams, how much does the Melbourne infestation weigh in metric tonnes? (A metric tonne is 1,000 kilograms.)

a) 456.80 metric tonnes
b) 102.55 metric tonnes
c) 19.19 metric tonnes

This is about five-eighths the weight of Bill Gates' entire fortune in single dollar bills.

And what do you call a 20-tonne colony of ants? Answer: ma'am, ma'am, ma'am, ma'am, ma'am, ma'am, ma'am ...

Biomassive

The total weight of all life on earth (the biomass) is estimated to be about 5 trillion metric tonnes (5 followed by 12 zeros). Staggeringly, ants are thought – in the highest academic estimates – to contribute perhaps 15 per cent of this total weight. Let's assume both these estimates are correct.

i. What, then, is the total weight of ants on earth?

a) 890 billion metric tonnes
b) 800 billion metric tonnes
c) 750 billion metric tonnes

Answers: ii) c i) c

An anteater can eat 30,000 ants in a day.

ii. Therefore if an 'average' ant weighs 2.5 milligrams, how many ants are there on earth?

a) 630 million trillion (630 quintillion - 63 followed by 19 zeros)
b) 300 million trillion (300 quintillion - 3 followed by 20 zeros)
c) 140 million trillion (140 quintillion - 14 followed by 19 zeros)

These are numbers that beggar belief, numbers difficult to feel at ease with. To help, consider this: it's estimated that the human body contains 50-100 trillion cells. The total number of ants is therefore equivalent to the number of cells in between 3 to 6 million people, for argument's sake the population of London.

But even further beyond comprehension is the highest estimate of the total number of insects on earth at 6 septillion (6 trillion trillion - 6 followed by 24 zeros). Let's assume it's true.

iii. Approximately how many insects would this mean there are for every one of the 6.388 billion humans on earth?

a) 1 billion billion (1 quintillion - 1 followed by 18 zeros)
b) 1 million billion (1 quadrillion - 1 followed by 15 zeros)
c) 1 million million (1 trillion - 1 followed by 12 zeros)

Which means the number of insects on earth right now is ten to twenty times the total number of human cells there are now, or half to as many human cells as there have ever been.

Answers: ii) b iii) b

Termites are thought to account for another 17 per cent of the biomass.

Antstronomy

300,000,000,000,000,000,000 – the size of the global ant column – is an astronomical number.

A light year is 9,460,528,405,000 kilometres. There are 1 million millimetres in a kilometre. The total number of millimetres in a light year is therefore 9,460,528,405,000,000,000. For ease, think of this number as 9.461 times 1 followed by 18 zeros.

Assume the average ant is 4 millimetres long. The total length of all the ants in the world, marching in a column antennae to tail, is therefore 12 followed by 20 zeros millimetres.

You can guess the question ...

i. How many light years would a column of all the ants on earth stretch?

a) 43.3
b) 88.1
c) 126.8

This is about as far as the star Tabit pi1 in the constellation of Orion, or the distance to the star Aldebaran in the constellation of Taurus and back.

And the city of Melbourne thinks it's got a problem!

Answer: i) c

In the Middle Ages having ants in your house was considered to be a lucky omen.

A.N.T. Please Call Home

Traversing the universe, if one of our insect travellers decides it wants to come home, it might be useful to have our accurate address. Fortunately the publishers of *Number Freaking* have agreed to provide a Poste Restante. In the way that school kids write their extended address inside the covers of their exercise books, our full address in the universe is:

Icon Books
Cambridge
Cambridgeshire
UK
Europe
Planet: Terra
Star: Sol
Galaxy: Milky Way – Orion Arm
Cluster: Local Group
Super Cluster: Virgo
This Universe

Because it'd be so easy to get lost in the vastness of space. Scaling stars down to the size of apples, our galaxy would shrink to be about the size of the inter-planetary distance between Earth and Venus. Individual stars, meanwhile, would, on average, be scattered apart as distant as Toronto is to Vancouver on the opposite coast of Canada.

Within our own solar system, if the sun were the size of an apple, our earth would be a grain of salt orbiting at a distance of about 10 metres. Jupiter would be the size of a blueberry about 60 metres out, and Pluto would be another salt grain 550 metres further on.

But regardless of any analogy about insects reaching the stars, it is, of course, impossible to comprehend astronomical distances. While most of us might just about be able to visualise the idea of a grain of salt orbiting an apple over half a kilometre away, nobody can 'see' others that are at least the width of a continent apart, let alone place them in the context of a starscape stretching from here to Venus.

But by Number Freaking it is possible to 'visualise' our own solar system. Imagine that the edge of the sun is the first word of a book - this book if you like - and the last word is the atmosphere surrounding the planet Pluto. The words and pages as you read are the miles between them passing by. On this scale - since the first word of this book - you are, right now, somewhere between Uranus and Neptune.

Solar Distance as Measured in Pages of a Book

Planet	Distance (in miles)	Distance (as pages of a 200-page novel)
Sun	0	cover
Mercury	35,983,100	1.96
Venus	67,237,900	3.66
Earth	92,955,800	5.06
Mars	141,635,700	7.71
Jupiter	483,634,000	26.32
Saturn	886,683,000	48.26
Uranus	1,783,951,000	97.1
Neptune	2,794,350,000	152.1
Pluto	3,674,490,000	200.01

(To include the newly discovered planet 'Sedna' on this scale - at 8 billion miles from the sun - this book would need 441 pages. Sedna would appear at page 440.52.)

Sedna was first discovered in March 2004; it is the first planet discovered in our solar system since Pluto in 1930.

Very Big and Very Small

In a scene towards the end of the 1998 animated movie *Antz*, the 'hero' (an ant named 'Z', voiced by Woody Allen) and the 'heroine' (an ant named 'Princess Bala', voiced by Sharon Stone) accidentally hitch a ride on the sole of a boy's trainer. To their ant-sized brains, the few yards they travel seem incomprehensibly vast; the time it takes, unbelievably fast.

The British garden ant, *Lasius niger*, is described as being 4 millimetres long. A man is, say, 1.75 metres tall.

> **i. If an ant were as big as a man, how tall would a man have to be to be the same relative size?**
>
> a) 765.63 metres
> b) 455.11 metres
> c) 222.44 metres

- *Empire State Building: 381 metres*
- *Eiffel Tower: 300 metres*
- *George Washington: 137 metres (George Washington's face carved on Mount Rushmore is 18 metres tall. If carved from head to toe at this scale, George would be 137 metres tall. Our 765-metre man's face would be 101 metres tall.)*
- *Big Ben: 96 metres*
- *Giant Sequoia (Redwood): 95 metres*

Assume a man 1.75 metres tall has a step length of 0.5 metres. Assume a normal walking pace is two steps per second. Our man therefore has a speed of 3.6 kph (2.24 mph).

Answer: i) a

ii. What would be the relative step length of a man 765.63 metres tall?

a) 159.66 metres
b) 176.54 metres
c) 218.75 metres

iii. Travelling at two steps a second, how fast would this be?

a) 867 kph
b) 1,202 kph
c) 1,575 kph

A running man could be seven times faster.

- *World's fastest fighter aircraft: MiG25 (maximum velocity 3,000 kph)*

Very Small and Very Very Small

Bacteria are by definition microscopic. Their size varies, but the most numerous species on earth (on land at least) is generally accepted to be a bacillus (rod shaped) bacterium called *Escherichia coli* (*E. coli*) which has an average length of around 5 micrometres (one 200th of a millimetre). It can cause various illnesses including gastroenteritis in humans.

i. If an *E. coli* bacillus were the size of an ant, what would be the relative size of an ant normally 4 millimetres long?

a) 3.8 metres
b) 3.4 metres
c) 3.2 metres

Answers: ii) c iii) c i) c

The earth orbits the sun at about 67,000 miles per hour.

- *Usual maximum length of a male Bengal tiger: 3.1 metres*
- *Usual maximum length of a female American alligator: 3.0 metres*

ii. On this scale how tall would a 1.75-metre man become?

a) 1,800 metres
b) 1,400 metres
c) 1,200 metres

- *Height of Ben Nevis: 1,392 metres*

So relatively speaking ... if a tummy bug were the size of an ant, an ant would be as big as a tiger and a man would be almost as tall as the highest mountain in Britain.

Bear this in mind next time you see a schlock horror movie in which people are attacked by giant insects.

Its Sister Causes Gangrene

The neurotoxin produced by the bacterium *Clostridium botulinum* is the most deadly poison on earth. While cosmetic surgeons use it as the active ingredient in Botox treatment, those nice people at the US Department of Defense have allegedly weaponised it for their friends in the army to play with ...

According to studies in primates, an injection of just 0.1 micrograms of the toxin is enough to kill an adult human, 0.75 micrograms will kill anyone unlucky enough to inhale it, and 75 micrograms will assure an agonising end to anybody who eats it. There are 1 million micrograms in a gram.

Answer: ii) b
While he was President Bill Clinton sent just two emails.

i. How much toxin would be required to kill all 6.388 billion people on earth by injection?

a) 0.6388 kilograms
b) 0.6982 kilograms
c) 0.7112 kilograms

(This is equivalent to about 1.41 pounds, or just under two-thirds of a 1 kilo bag, of flour.)

ii. How much toxin would be required to kill all 7,284,000 people in London by inhalation?

a) 7.34 grams (about a quarter of an ounce)
b) 5.46 grams (about a fifth of an ounce)
c) 14.67 grams (about half an ounce)

(Released from a radio-controlled model plane crashed into Nelson's Column perhaps?)

iii. How much toxin would be required to kill Britain's 59.66 million people by ingestion?

a) 46.34 kilograms
b) 19.78 kilograms
c) 4.47 kilograms

For everyone to eat it presumably it would need to be in something chocolatey.

Answers: i) a ii) b iii) c

Pest controllers estimate that there is one wild rat for every human in Britain.

What Goes Around Comes Around

The possibility that Asian avian flu might mutate into a human form and trigger an epidemic that could kill thousands if not millions has the World Health Organisation seriously rattled. There is already a global shortage of drugs that might combat such an outbreak.

The Worst Case Scenario is when not if ... because we're all up to our elbows in dead birds. Chicken has transcended the status of ingredient and become a component, mass-produced as efficiently as bottles or washers. Worldwide, humanity consumes around 1 billion chickens a week.

i. How many chickens is that per second?

a) 2,883
b) 1,653
c) 768

About 10,000 chickens are routinely slaughtered in the time it takes to read this sentence out loud.

- *A hen can lay 300 eggs in a lifetime.*

Eggs Is Eggs

People in the UK eat about 30 million eggs per day.

Answer: i) b

An average American eats 28 pigs in a lifetime.

i. How many is this per year?

a) 16,223,200,000
b) 10,957,500,000
c) 4,557,454,000

In two years the total number of people who have fallen ill due to *Salmonella* strains associated with eggs is 6,000. That's 3,000 per year.

ii. How many eggs are consumed in the UK for each case of *Salmonella* poisoning?

a) 5,191,122
b) 3,652,500
c) 1,355,670

Assuming the UK population is about 60 million, this means on average everyone in the UK eats one egg every two days.

iii. How many days, on average, would it take someone eating one egg every two days to be infected with *Salmonella* from them?

a) 13,908,000
b) 10,707,772
c) 7,305,000

iv. How many years is this?

a) 40,000
b) 20,000
c) 10,000

So how do you like yours? Boiled, scrambled or fried?

Answers: i) b ii) b iii) c iv) b

The world's biggest ever strawberry cake used 6,600 eggs.

Never Eat at a Place Called Mom's

Operation Desert Storm, the land campaign of what can now be called the First Gulf War, lasted around 100 hours. During that time the US Army Catering Division prides itself on having delivered over 2 million MREs – 'Meals Ready to Eat'. A logistical triumph, undoubtedly. According to veterans' reports, however, they were invariably cold – quite a feat itself, in a desert.

In this short space of time, under the ultimate command of the first President George Bush, a total of 47 British service personnel were killed, 22 of them by friendly fire.

There are 111,780 soldiers in the British Army including officers.

i. At this rate of attrition (47 men killed in 100 hours), about how long would the current British army last in an extended continuous conflict?

a) 6 years 15 weeks
b) 27 years 7 weeks
c) 43 years 34 weeks

ii. And how long under 'friendly fire' alone?

a) 12 years 23 weeks
b) 43 years 5 weeks
c) 57 years 50 weeks

When General George Armstrong Custer made his historic 'last stand' at the Little Big Horn in June 1876, the reality is, it wasn't much of a stand. A foolhardy Custer ordered his force of 197 troopers to attack what contemporary writers described as a temporary village of 10,000 Sioux (including perhaps 2,000–

Answers: i) b ii) c

3,000 fighters) stretching 8 kilometres along the Little Big Horn River in Montana. Frankly, what did Custer expect? The 7th Cavalry never made it past the Sioux's sentries. The engagement lasted, in the words of a Sioux witness, 'as long as it takes a hungry man to eat a hearty meal'.

In other words, it lasted about 20 minutes.

---

iii. At this rate of continuous attrition (197 men killed in 20 minutes), about how long would the entire British Army last under the command of a man like Custer?

a) 7 days 21 hours
b) 17 days 3 hours
c) 36 hours

So what can we conclude? Never go to war with a man named George.

Are We There Yet?

The average distance to the moon from the earth's surface is 232,841.69 miles (374,712.13 kilometres).

---

i. If there were a motorway route available to the moon, and you could drive your car at a constant speed of 70 mph (113 kph) – non-stop, without mishap – how long would it take you to reach the moon (in weeks [w], days [d], hours [h], minutes [m] and seconds [s])?

a) 19 w 5 d 14 h 18 m 35 s
b) 32 w 4 d 21 h 38 m 51 s
c) 7 w 2 d 8 h 44 m 36 s

Answers: iii) a i) a

Total bombs dropped in the Second World War: 2.7 million tonnes; in the Vietnam War: 6.73 million tonnes.

ii. More realistically, of course, there'd be traffic jams. Assuming that the average car is 4 metres long, how many cars would stretch nose to tail to the moon?

a) 93,678,032.5
b) 26,556,453.7
c) 47,128,212.6

Carmageddon

Estimates of the number of passenger cars in the world vary from 450 million to 710 million. We'll assume it to be 580 million (so enough to stretch to the moon and back 3.1 times). The average passenger car imported into the USA in 2001 had a kerb weight of 3,154 pounds. Assume this average holds true for every car. One kilogram is 2.205 pounds.

i. What is the total weight in kilograms of all the cars on earth?

a) 829.62 billion
b) 1.63 trillion
c) 2.99 trillion

Cold drawn steel has a specific gravity of 7.83. This means 7.83 grams of steel will occupy one cubic centimetre. Cars are built from various materials, but imagine they're built entirely of steel. Now imagine every car on earth were compressed into a single steel cube.

ii. How high would this cube be in metres?

a) 101.2
b) 473.2
c) 464.1

Which could fit the Empire State Building – at a shade under 381 metres tall – 1.24 times.

Dutch Estate Agents Are Best

In 1867 America paid Russia $7.2 million for its rights in Alaska. Alaska has a total area of 1,518,800 square kilometres.

i. How much did Alaska cost America per square kilometre?

a) $4.74 (1.92 cents an acre)
b) $11.56 (4.68 cents an acre)
c) $7.25 (2.93 cents an acre)

Meanwhile, in 1624, on another day in history, the Dutch bought Manhattan from the Algonquian native Americans for $24 in beads and other trinkets.

The island of Manhattan has an area of 87.5 square kilometres.

ii. How much did the Dutch pay per square kilometre of Manhattan Island?

a) 64.2 cents
b) 44.0 cents
c) 27.4 cents

Answers: ii) b i) a ii) c
Every day ten people in the UK die in road accidents.

iii. At this price how much would Alaska have cost?

a) $416,585.13
b) $798,663.76
c) $1,008,876.11

iv. Purchased at the same pro-rata price as the US paid for Alaska, how much would the island of Manhattan have cost?

a) $1,395.34
b) $950.10
c) $414.80

When America bought Alaska, maybe it should have used a Dutch realtor.

Is There a Family Discount?

Manhattan may have cost just $24 in 1624, but fast-forward 380-odd years, and at the time of writing, the rate for a 34-square-metre deluxe double room at New York's Plaza Hotel overlooking Central Park is now $439 per night.

i. At this rate ($439 for 34 square metres), how much would it cost to rent the entire area of Manhattan (87.5 square kilometres) for a night?

a) $1,129,779,350 ($12,911,764 per square kilometre; $12.91 per square metre)
b) $2,342,456,300 ($26,770,929 per square kilometre; $26.77 per square metre)
c) $3,873,009,100 ($44,262,961 per square kilometre; $44.26 per square metre)

Answers: iii) a iv) c i) a

The Statue of Liberty's mouth is one metre wide.

ii. How much would it cost for a year?

a) $293.55 billion
b) $345.91 billion
c) $412.65 billion

To put this number in perspective, in Fiscal Year 2004 the total United States defence budget was $399.1 billion: enough to rent the whole of Manhattan for 353 nights at Plaza Hotel room rates.

By comparison, the combined annual defence budgets of twenty of America's closest allies could rent Manhattan at Plaza room rates for only 349 nights.

Recent Annual Defence Budgets of Twenty Countries ($ billions)

Russia	65.0	Germany	24.9	Brazil	10.7	Canada	7.6
China	47.0	Saudi Arabia	21.3	Taiwan	10.7	Netherlands	6.6
Japan	42.6	Italy	19.4	Israel	10.6	Turkey	5.8
UK	38.4	India	15.6	Spain	8.4	Mexico	5.9
France	29.5	South Korea	14.1	Australia	7.6	Vietnam	2.4

Combined total of all the above: $394.1 billion

Sources: International Institute for Strategic Studies, US Department of Defense

Total military spending in the 'Axis of Evil', meanwhile, with a combined value of only $9.3 billion (just 2.3 per cent of the US total), would be sufficient to rent Manhattan at Plaza Hotel rates for only 8.2 nights – not much more than a short break.

Annual Defence Budgets in the Axis of Evil ($ billions)

Iran	4.8	North Korea	2.1	Syria	1.0	Cuba	0.8	Sudan	0.6

Sources: International Institute for Strategic Studies, US Department of Defense

(Rooms to be vacated by noon please.)

Answer: ii) c

Running on petrol an aircraft carrier would get 6 inches to the gallon.

Mark Twain Was Right

Money depreciates in value over time. Allowing for inflation, a dollar from 1867 was worth about $12.20 in 2004, and a dollar from 1624 at least $20.41 (the value of a dollar from 1665 – the year after the Brits seized Manhattan from the Dutch).

Alaska has a total area of 1,518,800 square kilometres.

> **i. In 1867 Alaska cost the USA $7.2 million. How much is this in modern dollars?**
>
> a) $125.9 million
> b) $34.8 million
> c) $87.8 million

> **ii. How much did Manhattan (at $24 in 1624) cost the Dutch in modern dollars?**
>
> a) $1,246.33
> b) $845.98
> c) $489.84

Buy land, my boy, they don't make it any more.

Landruff

An average person has 1.5 to 2 square metres of skin (call it 1.75) and will shed about 18.2 kilograms of dead skin in a lifetime.

Answers: i) c ii) c

The gold extracted from 1 tonne of ore is enough to make one wedding ring.

i. What is the combined skin area of everyone on earth?

a) 5,224 square kilometres
b) 11,179 square kilometres
c) 19,345 square kilometres

- *Area of Connecticut: 12,973 square kilometres*

ii. How many metric tonnes of dead skin will all the people on earth (6.388 billion) shed between them in a lifetime?

a) 67.34 million metric tonnes
b) 116.26 million metric tonnes
c) 243.22 million metric tonnes

- *Annual global fish consumption (by humans) 2003: 116 million metric tonnes*

iii. Assuming the average person weighs 70 kilograms, how many extra people is this equivalent to?

a) 3.84 billion
b) 1.66 billion
c) 693 million

Which is equivalent to 26 per cent of the existing global population.

Answers: i) b ii) b iii) b

In its 30-day lifetime a head louse will lay 100 eggs.

Padding

There's an urban myth about Miami expressed thus: 'One in five women in this town has had a boob job.'

There are 107,534 women aged between 18 and 64 resident in Miami.

> **i. How many 18–64-year-old women in Miami are rumoured to have had a boob job?**
>
> a) 17,345
> b) 21,507
> c) 35,893

The term 'boob job' can cover a number of cosmetic surgical procedures: breast reduction, breast reconstruction, breast augmentation (enlargement) and nipple realignment to name but four. Equally, while the majority of procedures will involve both breasts, a percentage will involve just one.

Another urban myth about Miami is that 70 per cent of all boob jobs carried out there are cosmetic augmentations involving prosthetic implants. Let's assume it's true. Let's also assume that these procedures invariably involve both breasts.

A modern breast implant might typically weigh 575 grams.

Answer: i) b

Plastic surgery dates back to 600 BC.

ii. Among women aged 18–64 what is the collective 'installed' weight of every cosmetic prosthetic breast in Miami?

a) 17,182 kilograms
b) 17,313 kilograms
c) 17,984 kilograms

So there are well over 17 tonnes of fake breasts in Miami.

Schoolyard

This is a British schoolyard number freak you may have come across. It supposedly proves you don't go to work. Here's the reason why:

Start with a leap year of 366 days ...

Assume you work from 9 a.m. to 5 p.m.

This means you work the equivalent of a third of all days, which is 122 days.

But you don't work weekends so subtract 104 days.

That leaves eighteen days.

Assume you get fourteen days' paid holiday. Subtract that too.

Four days left ...

Finally you don't work on at least four of the following days:

Christmas Day, Boxing Day, New Year's Day, Easter Sunday, May Day, Whitsun, August Bank Holiday, Thanksgiving, Labor Day, Passover, Eid.

Conclusion? You never work.

Answer: ii) b

Seventy-five per cent of women wear the wrong sized bra.

Star Value

Here is a list of celebrity salaries. Think of it as a list of prices.

Aerosmith	$25m	Keanu Reeves	$30m
Anna Kournikova	$10m	Lennox Lewis	$28m
Ben Affleck	$40m	Lisa Kudrow	$26m
Britney Spears	$39.2m	Madonna	$43m
Bruce Willis	$46m	Meg Ryan	$25m
Cameron Diaz	$40m	Mel Gibson	$40m
Christy Turlington	$6.2m	Michael Jordan	$36m
Courtney Cox-Arquette	$24m	Milla Jovovich	$4.3m
David Letterman	$32m	Nicole Kidman	$15m
Elton John	$30m	The Olsen Twins	$15m
Harrison Ford	$30m	Oprah Winfrey	$150m
Heidi Klum	$4.5m	The Osbournes	$1.6m
Howard Stern	$31m	Reese Witherspoon	$18m
Jay Leno	$20m	Robert DeNiro	$40m
Jennifer Aniston	$24m	Serena Williams	$8m
Jennifer Lopez	$37m	Stephen King	$52.4m
Julia Roberts	$20m		

i. Clearing out the back of a cupboard you find four old Elton Johns, one unopened Heidi Klum, and a pair of Reese Witherspoons. If you decide to swap them for a Bruce Willis and a Britney Spears, how many pairs of Olsen Twins can you still afford?

a) 5
b) 12
c) 26

Answer: i) a

On Valentine's Day 15 per cent of women send flowers to themselves.

ii. Those Osbournes, they get everywhere! And Aerosmith are looking a bit battered. And frankly, do you still need that Stephen King? If you swapped them all for a Keanu Reeves and a Harrison Ford, how much money would you have left to put towards a new Ben Affleck?

a) $29 million
b) $19 million
c) $9 million

iii. Solve this equation ...

$$\frac{\text{Cameron Diaz} + \text{Julia Roberts}}{\text{Jennifer Aniston} + \text{Lisa Kudrow}} = \frac{\text{Meg Ryan} + \text{Nicole Kidman} + ???}{\text{Mel Gibson} + \text{Anna Kournikova}}$$

a) Jennifer Lopez
b) 2 x Serena Williams
c) Jay Leno

Courtney Cox-Arquette will be appearing in the next episode of this Number Freak.

Just What Are the Other 56 Varieties?

Just over half the beans in the UK (51 per cent) are currently supplied by Heinz.

Relative Per Capita Consumption of Baked Beans		
Ireland	000000000000000000000000000	(27)
UK	00000000000000000000000	(23)
New Zealand	00000000000	(11)
USA	0000000000	(10)
Australia	000000000	(9)
Canada	000000	(6)

Answers: ii) b iii) c

Oprah Winfrey is worth about 94 times the Osbournes.

Take 229,000 tonnes of baked beans. Cover them in 59 million litres of tomato sauce. Share them between 696 million 415-gram cans 10.8 centimetres in height and 7.5 centimetres in diameter, and what do you get? The larder load of baked beans the British munch their way through every year.

i. What is the volume of a can of baked beans?

a) 592 cubic centimetres
b) 477 cubic centimetres
c) 355 cubic centimetres

There are a million cubic centimetres in a cubic metre.

ii. What is the total volume of 696 million cans of baked beans of this size?

a) 331,992 cubic metres
b) 438,887 cubic metres
c) 241,446 cubic metres

If all these tins of beans were crushed together into one giant cubic cardboard box it would have sides of nearly 70 metres and weigh 288,840 tonnes. The Great Pyramid at Giza has a volume of 2.1 million cubic metres.

iii. Presuming all the tins were hammered into neatly interlocking block shapes by teams of helpful metal workers, how many years' worth of the UK national baked bean supply would be required to construct a pyramid of baked bean cans the size of the Great Pyramid?

a) 11.23 years
b) 6.33 years
c) 2.42 years

Answers: i) b ii) a iii) b

- *The cans of beans consumed annually in America would build the Giza Pyramid in just under three years.*

Football Mad

These are the final scores in every Super Bowl.

XXXIX	24–21	XXVI	37–24	XIII	35–31
XXXVIII	32–29	XXV	20–19	XII	27–10
XXXVII	48–21	XXIV	55–10	XI	32–14
XXXVI	20–17	XXIII	20–16	X	21–17
XXXV	34–7	XXII	42–10	IX	16–6
XXXIV	23–16	XXI	39–20	VIII	24–6
XXXIII	34–19	XX	46–10	VII	14–7
XXXII	31–24	XIX	38–16	VI	24–3
XXXI	35–21	XVIII	38–9	V	16–13
XXX	27–17	XVII	27–17	IV	23–7
XXIX	49–26	XVI	26–21	III	16–7
XXVIII	30–13	XV	27–10	II	33–14
XXVII	52–17	XIV	31–19	I	35–10

i. What is the average score in a Super Bowl Final (approximately)?

a) 34-19
b) 31-15
c) 26-11

And no finalists have ever scored it.

Answer: i) b

An average American eats 87 hot dogs per year.

Planet IKEA

Population Density (people per square kilometre)

Monaco	16,000	India	318	Ireland	55
Singapore	6,430	UK	244	Afghanistan	43
Malta	1,260	China	134	USA	29
Netherlands	393	France	109	Australia	2.5
Japan	336	Spain	79	Sahara	1.0

The total surface area of land on earth is 148,327,070 square kilometres. The total global population is 6.388 billion.

i. How many people is that per square kilometre of land?

a) 23.983
b) 43.067
c) 65.293

- *Population density in Afghanistan (people per square kilometre): 43*

ii. And how much land is there per person on earth?

a) 1.434 hectares (14,340 square metres)
b) 2.322 hectares (23,220 square metres)
c) 3.748 hectares (37,480 square metres)

So how big is this? Think of a square with sides of 500 feet (499.94 to be exact).

Answers: i) b ii) b

In the 1990s, the US population increased by over 25 million.

Alternatively, at the time of writing, there are 186 IKEA stores worldwide. Their combined floor area totals 3,979,600 square metres.

> ### iii. What is the floor area of the average IKEA store?
>
> a) 15,123.4 square metres
> b) 21,395.7 square metres
> c) 26,876.1 square metres

The earth's land surface divides up into about a plot the size of an IKEA store for everyone.

This Island Earth

The world record for the number of people who can simultaneously squeeze into a British red telephone kiosk is twelve. The classic K6 red telephone kiosk has a floor area of 0.836 square metres. So that's fourteen people per square metre.

> ### i. At fourteen people per square metre, how big an area would you need to fit the entire global population of 6.388 billion?
>
> a) 433.6 square kilometres
> b) 447.2 square kilometres
> c) 456.3 square kilometres

- *Area of the Principality of Andorra: 453 square kilometres*

Historically British schoolchildren were taught that the world's population could fit on the Isle of Wight – which has an area of 380.99 square kilometres. While this was once true, it hasn't been for a long time. At fourteen people per

Answers: iii) b   i) c

3,979,600 square metres is equivalent to 577 premiership football pitches.

square metre it wouldn't have been possible much after 1990, and at a more realistic five people per square metre not since the end of the First World War.

But imagine if everyone on earth now (6.388 billion) were packed in shoulder to shoulder, like penguins, stood upright for maximum space at, say, five people per square metre.

> **ii. How big an area would you need to cram everyone on earth together?**
>
> a) 1,443.3 square kilometres
> b) 1,332.5 square kilometres
> c) 1,277.6 square kilometres

- *Area of the Isle of Skye: 1,666 square kilometres*
- *Area of the demilitarised zone between North and South Korea: 1,262 square kilometres*
- *Area of Sao Paulo, Brazil: 1,168 square kilometres*

But would everybody in the world want to fit in between North and South Korea?

Close Neighbours

According to the UN, 48 per cent of people now live in cities. (By 2007 it will be the majority.) Currently the most densely populated city on earth is Jakarta, Indonesia, which squeezes 50,203 people into every square kilometre. That's just under 20 square metres each. So how big is that? Think of a square with sides just under 15 feet. Alternatively think of five double beds pushed together ...

Answer: ii) c

Four prisoners per square metre were crammed into the Black Hole of Calcutta.

i. Crammed together as tightly as in Jakarta, how big an area would you need to pack in everyone on earth?

a) 127,243 square kilometres
b) 334,335 square kilometres
c) 534,224 square kilometres

- *Area of Louisiana: 125,679 square kilometres*
- *Area of New York State: 128,401 square kilometres*
- *Area of England: 130,477 square kilometres*

(In reality people in Jakarta increase their space by living in multi-storey dwellings.)

ii. If the average Jakartan gets 20 square metres, how many Jakartans could live in the average 21,395.7 square metre IKEA store?

a) 1,069.8
b) 1,442.5
c) 1,823.8

At least they'd all find a bed. And a kitchen. And various odd-shaped bits of plastic that defy explanation.

Hell Isn't Other People

Living in the crush of Jakarta, you might consider it hellish if you weren't used to it. You might consider it hellish even if you were. But is it anything like the real thing?

Answers: i) a ii) a

The US Census Bureau defines a town as having up to 2,500 residents, and a city 2,501 or more.

Let's consult an expert.

In Dante's 14th-century masterpiece *The Divine Comedy*, Hell extends through nine circles from the earth's surface to its core – where we are told Satan sits alone encased in ice. Assume each circle contains the same number of sinners as the next (which, statistically, one would expect over time).

The total number of people who've ever lived is estimated to be between 96 and 110 billion, including the 6 billion plus people currently alive. The number of people who've ever died therefore ranges from 90 to 104 billion. Assume it to be 97 billion. And they've all gone to Hell.

If 97 billion people have ever died, each of Dante's vast terraces must now contain some 10.78 billion souls (over two-thirds more than the current global population) and must, particularly as one gets closer to the centre, be getting very crowded and uncomfortable indeed.

Or so you'd have thought.

The equatorial radius of the earth is 6,378.14 kilometres. Each circle of Dante's Hell could be over 600 kilometres deep. Not exactly a low ceiling.

This table number freaks nine circles descending onion-like between the earth's surface and core, and the space each sinner could claim.

Circle of Hell	Total Area (square kilometres)	Area Per Sinner (square metres)[1]	(square metres)[2]
1	414,245,421	38,435	10,765
2	327,305,024	30,368	8,506
3	250,592,909	23,250	6,512
4	184,109,076	17,082	4,784
5	127,853,525	11,862	3,322
6	81,826,256	7,592	2,126
7	46,027,269	4,270	1,196
8	20,456,564	1,898	531
9	5,114,141	474	133

[1] Assuming there are no oceans of any type in Hell
[2] Assuming the 'surface' of Hell is 72 per cent larva oceans

Ironically it suggests that, due to a lack of oceans, in the outer circles there's more room per sinner in Hell than in life.

Jakartans make do with 20 square metres of space each. Even in the deepest circle of Hell sinners still have more space than the average Jakartan.

Shopping Is Hell

According to IKEA 310 million people visit their 186 stores in 31 countries every year.

i. On average how many people visit each IKEA store per day?

a) 2,399
b) 4,563
c) 8,772

Allowing for bank holidays, etc., the IKEA store in Croydon, UK, is open on average about 12.6 hours a day. Assume this is a global norm.

ii. About how many visitors does that average out at per hour per store?

a) 362
b) 776
c) 1,223

Assume the average visitor stays at IKEA for around three hours.

Answers: i) b ii) a

The Victorian census defined a family with two or fewer servants as 'lower middle class'.

iii. So on average how many visitors are in an average IKEA store at any one time while it's open?

a) 2,443
b) 557
c) 1,086

So an afternoon at IKEA is like visiting Jakarta, the most crowded city on earth. But be warned: it's more crowded than Hell.

The Human Swarm

According to the *CIA World Factbook*, 106 boys are born worldwide for every 100 girls. Among adults over 15 this ratio falls to 103 men for every 100 women.

Twenty Places Where Adult Women Outnumber Men

Brazil	Estonia	Burma	Thailand
Colombia	Hungary	Cambodia	Singapore
South Africa	Latvia	Puerto Rico	Russia
El Salvador	Monaco	Nicaragua	Bolivia
Poland	Bulgaria	Portugal	Bahamas

There are 4.472 billion people aged over 15 on earth. Assume there are therefore 2.269 billion men and 2.203 billion women aged over 15 worldwide. (Global life expectancy for men is 65, for women 69.)

Answer: iii) c

There are a total of 364 presents in the song 'The Twelve Days of Christmas'.

Assume that worldwide an average man weighs 70 kilograms and an average woman 60 kilograms.

i. How many tonnes do all the people on earth aged over 15 weigh?

a) 1.344 billion
b) 291.01 million
c) 63 million

Adult humans (aged over 15) have an average body density (specific gravity) of 1.03 (exhaled). This means 1.03 tonnes of human would occupy 1 cubic metre of space.

ii. How many cubic metres of people aged over 15 are there?

a) 1.305 billion
b) 282.53 million
c) 61.17 million

So if everyone over 15 ever had the misfortune to be homogenised into one amorphous blob ...

iii. How high, in metres, would a cubic box to pour them all into have to be?

a) 1,092.8
b) 656.18
c) 294.02

(... which is more than twice the height of the Eiffel Tower.)

- *Height of the Taipei 101 building (tallest building on earth): 508 metres*
- *Height of Crowden Head, Kinder Scout (highest point in the UK Peak District): 631 metres*

Answers: i) b ii) b iii) b

A can of cat food contains as much meat as five adult mice.

Top Muscle

There are 639 named muscles in the human anatomy. Your doctor is supposed to know the names of all of them. But you don't need to be a doctor to know that one of the most important is the heart.

On average an adult human heart weighs 300 grams. The global population is 6.388 billion. Seventy per cent of the world's population are adults.

> i. What is the combined weight, in metric tonnes, of every adult human heart beating on the planet right now?
>
> a) 1.34 million metric tonnes
> b) 1.85 million metric tonnes
> c) 2.34 million metric tonnes

And all of them are broken.

My Lourdes, Ladies and Gentlemen

In the 19th century the Catholic Church established a medical bureau near the French shrine of Lourdes to authenticate, but usually deny, reports of miracles.

In the 120 years since it opened for business the bureau has received between 6,000 and 7,000 claims. We'll call it 6,500. Why the number is uncertain is not clear. Presumably if your sight has been miraculously restored you don't change your mind the next day. So it must be bad record keeping. Either that or God moving in mysterious ways.

Answer: i) a

There are 206 named human bones.

i. How many miracles have been reported on average every year?

a) 86
b) 20
c) 54

That's about one a week.

Because there's always one, isn't there? On any package tour there's always one person who can drink more than anyone else, always one person who's been more places than anyone else, always one person who has more adventures than anyone else. Presumably most trips to Lourdes are package tours a week long – and so presumably, there's always one person who's had more miracles than anyone else ...

In fact, of the miracles reported to the medical bureau, a total of only 66 – about 1 per cent – have been 'accepted'. So about 99 per cent have been rejected. Only one claim has been accepted in the 32 years (384 months) since 1972.

ii. On average how often were miracle claims accepted prior to 1972?

a) once every 11.96 months
b) once every 16.25 months
c) once every 22.44 months

You were about 24 times more likely to get a miracle accepted in the old days.

Answers: i) c ii) b

Two thousand seven hundred police officers were needed to protect Pope John Paul II the last time he visited Lourdes.

Soulitaire

Twenty-one grams is, famously, the apocryphal weight of the human soul.

> i. Given a global population of 6.388 billion, what, theoretically, is the combined weight in metric tonnes of the souls of everyone alive on earth today? (A metric tonne is 1,000 kilograms.)
>
> a) 103,445 metric tonnes
> b) 134,148 metric tonnes
> c) 166,877 metric tonnes

Global Population Change (2004)

Period	Births	Deaths	Net Increase
Year	129,108,390	56,540,896	72,567,494
Month	10,759,033	4,711,741	6,047,291
Day	352,755	154,483	198,272
Hour	14,698	6,437	8,261
Minute	245	107	138
Second	4.1	1.8	2.3

Figures may not add to totals due to rounding. Source: US Census Bureau

Two sentences:

1. 'Four people die every two seconds.'
2. 'A dozen babies are born every three seconds.'

Say either one out loud and it's done.

Answer: i) b

Twenty per cent of Americans are cremated.

ii. If we count only the souls of the dead, what weight of souls (in metric tonnes) is being 'released' every day? (A metric tonne is 1 million grams.)

a) 15.69 metric tonnes
b) 8.66 metric tonnes
c) 3.24 metric tonnes

iii. And what weight of souls (in metric tonnes) passes on every year?

a) 594.00 metric tonnes
b) 1,187.36 metric tonnes
c) 2,323.23 metric tonnes

Don't Pay the Ferryman

Global Populations throughout History (millions)

Year	World	Europe	North America	UK	London
-10000	8				
-3000	100	2			
-2000	200				
0	300	31		0.6	0.06
1650	510	103	0.5	7.5	0.4
1700	652	122	1.2	9.25	0.6
1750	795	144	3	10	0.65
1800	969	185	5.3	16	1.1
1850	1,265	227	23.2	25	2.65
1900	1,656	450	76	38	6.5
1950	2,513	549	151	50.23	8.2
2000	6,199	678	420	58.64	7.17

The figures for North America include Canada and Mexico. Sources: various

Answers: ii) c iii) b

The Google search engine yields over 20 million entries for the word 'Soul'.

If 97 billion people have ever lived ...

---

i. What is the total weight of all the 21-gram eternal souls there have ever been?

a) 2.04 million metric tonnes
b) 2.56 million metric tonnes
c) 3.05 million metric tonnes

- *Annual amount of trash produced by the city of San Diego: 2 million tonnes*

In the afterlife? The floor must take some battering.

Great Dancers

---

i. How many angels can dance on the head of a pin?

a) 5
b) 22
c) 1,643

- *Previous answers include 'countless' and 'ten septillion'.*

Heaven

According to the child-friendly version, Heaven is a kind of spiritual strato-sphere surrounding the entire earth. Number Freaking isn't a tool to prove it exists, but it can predict how big it might be.

Answers: i) a i) a
Fifty thousand bees weigh about 5 kilograms.

Dividing all the world's land between the entire global population works out at 23,220 square metres each. In the afterlife, where we're assured things will be better, each of us will presumably get even more ... a minimum, say, of 23,221 square metres. We'll also assume the Jesuits are right and we all go to Heaven ...

i. If all 97 billion ex-humans are in fact in Heaven, what is its minimum area?

a) 5,223,667,000 square kilometres
b) 2,252,437,000 square kilometres
c) 698,223,000 square kilometres

Land actually makes up only 29.08 per cent of the earth's surface; assume the same ratio in Heaven.

ii. What, then, is the minimum size of Heaven?

a) 25,992,765,300 square kilometres
b) 16,876,113,900 square kilometres
c) 7,745,656,800 square kilometres

A sphere with a surface area this big has a radius of 24,827 kilometres. The equatorial radius of the earth is about 6,378 kilometres.

iii. What is the minimum distance above earth that Heaven could therefore be located?

a) 42,556 kilometres
b) 30,996 kilometres
c) 18,449 kilometres

Answers: i) b ii) c iii) c

In the mid-14th century the Black Death killed over 20 million people in Europe.

- *Orbital height of geosynchronous communications satellites: 35,786 kilometres*

So we've been looking down rather than up at Heaven since the first geo-synchronous satellite in 1963.

Actually, all this is almost certainly an underestimate: it's based solely on the number of people already dead; if Heaven exists there's a lot more souls will be arriving yet ...

Global Population Growth

Year global population reached ...

1 billion: 1804

2 billion: 1927 (123 years later)

3 billion: 1960 (33 years later)

4 billion: 1974 (14 years later)

5 billion: 1987 (13 years later)

6 billion: 1998 (11 years later)

Year global population expected to reach ...

7 billion: 2009 (11 years later)

8 billion: 2021 (12 years later)

9 billion: 2035 (14 years later)

10 billion: 2054 (19 years later)

11 billion: 2093 (39 years later)

The UN project population will stabilise at 9 billion.

Source: UN Population Division

Maybe Heaven expands as it needs to.

On average, US funeral homes each process one corpse every two days.

Divine Works

So what do we know so far? Heaven is huge, there are loads of people in it, and since it's been around for some while, its carpet must be looking a bit threadbare.

Which sounds like a golden opportunity for the UK's carpet retailers. Particularly given how the juggernaut of laminate flooring has decimated the nation's carpet market.

UK Carpet Market

Year	Value (£ million)	Area (square metres)
2002	1,337	243.1 million
2003	1,319	236.0 million
2004	1,313	230.0 million

i. On average how much did a square metre of carpet cost in the UK in 2004?

a) £5.71 ($10.26)
b) £17.13 ($30.78)
c) £24.36 ($43.77)

ii. At this price how much would it cost to re-carpet Heaven's 2,252,437,000 square kilometres?

a) £8,349,111,000,000,000 ($15,000 trillion)
b) £12,861,415,000,000,000 ($23,000 trillion)
c) £20,037,614,000,000,000 ($36,000 trillion)

Answers: i) a ii) b

There are 5.5 ounces of salt per gallon of sea water.

No wonder the market for laminate flooring is up. To finance the carpeting of Heaven would take all of global GDP at current rates for 632 years plus.

The Shag Pile Index (SPI)

In a noteworthy act of Number Freaking, the financial magazine the *Economist* has created what it calls the Big Mac Index, which compares and contrasts the price of a McDonald's Big Mac around the world. So, for example, at the time of writing and at current exchange rates, a Big Mac which costs $2.90 in the US costs the equivalent of $3.37 in the UK, and $2.33 in Japan. The point of it is that by the application of some advanced economic jiggery-pokery the index apparently (but, by the *Economist*'s own admission, spuriously) allows monetary propeller-heads to estimate whether different currencies are undervalued or overvalued against the US dollar.

It certainly sounds like an idea. Could the cost of carpeting create a new economic index of some sort? You decide.

i. Assuming an average UK price of £5.71 ($10.26) per square metre, how much would it cost to carpet all the land on earth (148,327,070 square kilometres)?

a) £342,893,274,000,000 ($616.23 trillion)
b) £846,947,560,000,000 ($1,521.72 trillion; $1.5 quadrillion)
c) £610,332,449,000,000 ($1,095.93 trillion; $1 quadrillion)

So at a current global GDP of $36.36 trillion, to carpet all the land on earth would consume the next 41.85 years' worth of the entire global economy.

Answer: i) b

A square metre of single dollar bills would be worth about $96.72.

Country	Total area (square kilometres)	Cost $m at UK 2004 prices to carpet country	GDP $m	'SPI' (years of GDP to meet cost to carpet)
Singapore	693	7,110.18	91,342	0.0778 (28.43 days)
Netherlands	41,526	426,056.76	511,556	0.8329 (304 days)
Malta	316	3,242.2	3,870	0.8377 (306 days)
Japan	377,833	3,876,566.5	4,326,444	0.8960 (327 days)
UK	244,030	2,503,748	1,794,858	1.3950
France	547,030	5,612,528	1,747,973	3.2109
Ireland	70,280	721,073	148,553	4.8540
Spain	504,782	5,179,063.3	836,100	6.1943
USA	9,629,091	98,794,473	10,881,609	9.0790
India	3,287,590	33,730,673	598,966	56.3148
China	9,596,960	98,464,809	1,409,852	69.8405
Australia	7,686,850	78,867,081	518,382	152.1409

These insights Number Freaking freely donates to the world's economists. See how the Shag Pile Index demonstrates unequivocally that to carpet the UK might take over a year (assuming a fitter could be found) but to carpet the whole of France might take more than three. Quod Erratum Demonstrandum.

One Hundred Per Cent Proof

It's said that at any given moment 0.7 per cent of the world's population is drunk. That's 44.7 million people: enough people to create the world's 26th most populous country; 2 million more than the population of South Africa; 4 million more than the population of Spain. But does it mean 0.7 per cent of the UK's 59.6 million people too?

On average, people in the UK each drink the equivalent of 8.4 litres (1.85 UK gallons) of pure alcohol every year. A half-litre (500-millilitre) bottle of lager, for example, contains maybe 19 millilitres of pure alcohol (3.8 per cent), a bottle of wine maybe 82 millilitres (11 per cent). It all adds up.

> **i. Averaging these 8.4 litres out, if people in the UK drank the same amount of pure alcohol every day, how much would they consume daily?**
>
> a) 19 millilitres
> b) 23 millilitres
> c) 25 millilitres

To become drunk you have to drink a certain amount at a certain pace. Interestingly, perhaps, your thumbs are where you feel drunk first. But in men, at least, even consumed quickly, a daily quota of pure alcohol this size wouldn't normally raise someone's blood levels high enough to make them legally drunk.

The magic number in terms of enough on average to make you legally drunk every day if you drink it quickly is about 11 litres of pure alcohol a year.

Volume of Pure Alcohol Drunk Per Person Per Year

Rank	Country	Litres	Rank	Country	Litres
1	Luxembourg	12.3	9	Denmark	9.5
2	Romania	12.1	15	UK	8.4
3	Portugal	11.7	17	Russia	8.1
4	Ireland	10.8	19	Australia	7.8
5	Czech Republic	10.6	26	USA	6.7
6=	France	10.5	27	Canada	6.6
6=	Germany	10.5	28	Japan	6.5
8	Spain	10.0	30	Bulgaria	6.2

Answer: i) b

(Gourmet human cannibals please note: a Luxembourgoise liver is twice as well marinated as a Bulgarian.)

Relaxed as a Newt

Surveys into drinking tend to focus on those aged 18 to 64. Other age groups are presumably too young or too broke. In the UK there are 17.93 million men in this age range, and 17.57 million women (35.5 million in total).

But 8 per cent of 18–64-year-old men in the UK don't drink alcohol. Thirteen per cent of women don't either. Thus 16.5 million men and 15.29 million women drink alcohol.

Using bingeing frequency (how often people have five or more drinks) as a measure of how often people get drunk, men get drunk 47 times a year and women 16. This 'when' is heavily skewed. Clearly at times like New Year many more people will be drunk than on most other days. But averaging the numbers equally across the year suggests that a grand total of nearly 2.8 million people get drunk in the UK every day: 2.12 million men and 670,000 women. This is a ratio of about 3 to 1. Doesn't this feel about right? Check it out yourself in your local bar.

Now, assume the sober-drunk-sober cycle takes 3.5 hours.

i. Dividing the day into seven periods of (almost) 3.5 hours, and dividing our 2.8 million 18–64-year-old drunks evenly between them, about how many 18–64-year-old drunks are there afoot in the UK at any one time?

a) 400,000
b) 178,000
c) 622,000

Answer: i) a

One unit of alcohol is 10 millilitres of pure alcohol. The human liver can metabolise one unit in about one hour.

Rounded up this is 0.7 per cent of the UK's population. So yes, it's true in the UK too: at any given time, on average, 0.7 per cent of the British population is drunk.

- *0.7 per cent of the UK's population are employed directly or indirectly in sport.*
- *0.7 per cent of the UK's population live in the Bournemouth and Poole area.*

Best advice? Keep away, no, run away from Bournemouth.

Cold Hard Cash Mountain

At the time of writing, according to the Bill Gates Net Worth Page, Bill Gates' personal fortune - based on Microsoft's stock price - stands at $31,004,950,033.64.

i. One paper US dollar bill weighs 1 gram. So how much would Bill's personal fortune weigh, in metric tonnes, in single dollar bills?

a) 45,883
b) 62,002
c) 31,005

ii. The weight of an average man in the UK is now 79.2 kilograms, and an average woman 65 kilograms. What, then, is the combined body weight, in metric tonnes, of all the approximately 300,000 18-64-year-old men and 100,000 18-64-year-old women drunk in the UK right now?

a) 15,345
b) 60,456
c) 30,260

Answers: i) c ii) c

In the 2001 UK Census 0.7 per cent of people identified their religion as Jedi.

So there you have it. At this moment Bill Gates' wallet weighs more than every drunk in Britain.

Alternatively Bill could afford 435 pints (about a barrel and a half) of beer at every one of the UK's 65,000 pubs every day for a year, and still have plenty of cash left for a curry and his cab fare home.

Or, at £1.50 a drink, Bill could afford to buy a few beers and a packet of crisps for every one of the 6.388 billion people on earth.

Super Super Rich

There's a long and fine tradition of philanthropy among the rich; a pheno-menon only a cynic would characterise in words like camel, eye and needle.

According to figures released by the US tax office, the 400 highest earners in America donate, per annum, an average of $25 million each to charity – the 400 people, we shall assume, that *Forbes* magazine reports as having a combined net worth of $955 billion.

> ### i. What percentage of their total net worth do America's richest 400 donate to charity each year?
>
> a) 1.05 per cent ($10 billion)
> b) 2.15 per cent ($20 billion)
> c) 3.15 per cent ($15.2 billion)

- *GDP Trinidad and Tobago: $10.2 billion*

According to *Forbes*, the net wealth of these 400 individuals had actually increased by 10 per cent – that's $86.8 billion – on the previous year.

Answer: i) a

The combined body weight of every dipso in the UK on an average day is 172,298 metric tonnes.

ii. Given that this $86.8 billion could therefore be defined as annual net income, what percentage of their total annual net income do America's richest 400 donate to charity each year?

a) 15.2 per cent
b) 11.5 per cent
c) 9.3 per cent

The average UK salary is £25,170. Assume that net after tax it's 75 per cent of this – i.e. £18,878. At a rate of 11.5 per cent this is equivalent to an average British wage earner donating £2,171 to charity a year.

Two-thirds of the UK population donated to charity in 2002. The average total donation was about £155.

How to Make a Camel Smoothie

One of the better-known rules about going to Heaven is the one about a camel passing more easily through the eye of a needle than a rich man getting in. So the question is, how do you get a camel through the eye of a needle? Simple answer: use a blender.

A dromedary camel weighs 450–650 kilograms, so we'll assume 550 kilograms. In an average 79.2-kilogram man, the skeleton accounts for about 15 per cent of body weight. We'll assume the same is true for a camel. This means a camel contains about 80 kilograms of bone, and 470 kilograms of flesh. Human flesh has an average density of about 1.03 kilograms per litre, and human bone about 1.85 kilograms per litre. Again we'll assume these numbers work for our hypothetical camel too.

Answer: ii) b
The total amount given to charities by individuals in 2002 in the UK was £7.3 billion.

> **i. If you blend and blend and blend a camel down to liquid, with no waste or spillage, about how many litres will you be left with?**
>
> a) 335 litres
> b) 419 litres
> c) 500 litres

So how big is the eye of a needle? It depends on the size of the needle. Here are the dimensions of the eye of one modern steel sewing needle: 1 millimetre long, 0.3 millimetres wide, 0.5 millimetres deep – a total volume of 0.15 cubic millimetres.

> **ii. Having blended our camel, it must be pipetted through the eye of our needle. How many individual 'doses' will it take to pipette it through a needle eye with a volume of 0.15 cubic millimetres? There are 1 million cubic millimetres in a litre.**
>
> a) 5,361,963,554
> b) 4,009,227,003
> c) 3,333,333,333

Now clearly this would be a tedious job, so let's automate it with an automated pipetting machine that can deliver each dose at a rate of, say, once every 1.5 seconds.

> **iii. How many seconds will it take to pipette our camel through our needle?**
>
> a) 6 billion
> b) 4 billion
> c) 5 billion

Answers: i) c ii) c iii) c

iv. How long is that in years and days?

a) 387 years 48 days (387.13 years)
b) 158 years 161 days (158.44 years)
c) 173 years 12 days (173.03 years)

This would mean no rich man who's died since 1847 (which includes Rockefeller, Vanderbilt, Astor and Carnegie) has been able to enter Paradise yet. If it is a place with a bevy of virgins waiting, they must be getting very bored. But Number Freaking isn't science. Here are the dimensions of the 'eye' of another needle: 5 mm x 0.7 mm x 0.5 mm (1.75 cubic millimetres). Pipetting our camel through this hole would take only about 13.6 years - equivalent to the time since 1992. For the chance of an eternity in Paradise surely this wouldn't seem long at all. So which needle do you think the rich will have to pass through? Presumably it depends on their lawyers ...

Everyone Needs a Hobby

Of course there's also a long and fine tradition of the wealthy financing the military. Elias Howe (born 1846), the man who registered the first American patent on a sewing machine, amassed a fortune of nigh on $2 million between 1854 and his death in 1867 - an amount equal to $25 million in today's money. So wealthy was Howe, he equipped the entire Union Army infantry regiment in which he served as a private during the American Civil War, out of his own pocket ...

According to *Forbes* magazine the combined net worth of the wealthiest 400 citizens in the USA now is $955 billion - a 10 per cent ($86.8 billion) increase on last year.

Answer: iv) b

A camel can drink 25 gallons of water in 30 minutes.

- *Combined GDP of the world's poorest 85 nations: $970.5 billion*
- *GDP of Canada: $856.1 billion*
- *GDP of Singapore: $93.8 billion*
- *GDP of New Zealand: $76.9 billion*

Imagine they all agreed to continue the tradition of privately financing the military ...

The Cost of Some American Wars		
Conflict	Cost at time	Cost in 2004 dollars
First World War	$33 billion	$602 billion
Second World War	$360 billion	$4.92 trillion
Korean War	$50 billion	$417 billion
Vietnam War	$111 billion	$596 billion
First Gulf War	$61 billion	$83 billion
Grenada invasion, 1983	$76 million	$143 million
Panama invasion, 1990	$163 million	$234 million

At current prices America's richest 400 could have privately financed the First World War, the war in Korea, or Vietnam. With a bit of cost cutting (or maybe just another year's income) they could have afforded Korea and Vietnam combined. The First Gulf War they could have paid for out of their $86.6 billion annual income. As $86.6 billion is roughly $237 million a day, they could afford a Grenada or a Panama daily.

Bill Clinton once claimed the war in Afghanistan cost $1 billion a month ...

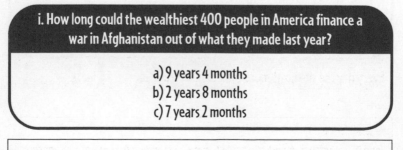

i. How long could the wealthiest 400 people in America finance a war in Afghanistan out of what they made last year?

a) 9 years 4 months
b) 2 years 8 months
c) 7 years 2 months

Answer: i) c

ii. And for how long could the wealthiest 400 people in America finance a war in Afghanistan if they spent their entire $955 billion?

> a) 82 years and 11 months
> b) 54 years and 8 months
> c) 79 years and 7 months

Which, given that the average US life expectancy is 77 years, is more than a lifetime ...

Greetings from the Global Yard Sale

Just as the sewing machine changed 19th-century lives, so the internet has changed ours.

eBay was launched in September 1995. Operating worldwide, every day 1 million new items are put up for sale. The site offers goods for sale in over 9,000 categories.

i. On average, how many new items are put up for sale every second?

> a) 69.4
> b) 105.6
> c) 11.6

Every day the site hosts more than 10 million auctions.

Answers: ii) c i) c

John D. Rockefeller, America's first billionaire, was at one time worth 1.53 per cent of the US economy.

Pussy Galore

Ian Fleming found the idea for James Bond's sobriquet '007' on documents prepared for the Queen by Dr John Dee, spymaster to Elizabeth I. An alchemist in his spare time, Dee used the double zero, even in the 16th century, to mean 'For Your Eyes Only' and the seven because he believed it to be mystical.

In his first twenty movies James Bond made love 79 times: mainly to brunettes and mainly in hotels. Assuming each of these movies is 2 hours long, in the 40 hours we have shared Bond's world, he has physically imposed himself on anything in a skirt about once every 30 minutes.

In fact there were 58 women, so sometimes James went back for seconds. Assume, therefore, that his liaisons per woman have lasted about 41.38 minutes. What a gent.

Notwithstanding the inherent risks of a job like Bond's, a British man has a life expectancy of 77 years. We shall assume a man of Bond's psychopathy loses his virginity at the age of 16, giving him an unabating predatory life of 61 years.

i. In continuous liaisons of 41.38 minutes each, how many women would James pass through in 61 years?

a) 775,340
b) 203,499
c) 621,865

Amateur.

(You can decide for yourself whether the women whose 41.38 minutes passed while he was asleep were lucky or not.)

Answer: i) a

By the same measure James would be told he would die 376,970 times.

Lost in Space

Many issues perplex the space exploration community. One such is the threat posed by space junk: defunct satellites, spent rocket stages, sheered bolts, mislaid tools, discarded food pouches and abandoned waste bags, broken pens and bent CDs; the detritus of clumsy, untidy or just unlucky space ventures, which might one day threaten the safety of the astronauts of the future.

Space scientists have predicted that around 100,000 fragments of space rubbish will eventually be detected by the time a definitive catalogue is completed. None of it can be sold on eBay.

Usually, objects orbiting the earth eventually do one of two things: either they drift off into the void, or they fall back through the atmosphere in a hail of sparks. The only exceptions are objects in what is known as geosynchronous orbit. Space agencies already exploit the physics of geosynchronicity to position communications satellites for maximum longevity.

Geosynchronous orbit is the crossover point between orbits inevitably destined for infinity and those destined for incineration – where orbital decay exactly matches the curvature of the earth; an endless free fall in concert with the surface below. But the key thing to remember is that geosynchronous orbit is the only orbital distance where the flotsam and jetsam of the space race really pose any kind of long-term threat.

So now let us make, for argument's sake, the unlikely assumption that all the putative 100,000 pieces of space junk actually are in geosynchronous orbit.

Scientists know how high geosynchronous orbit is: it's 35,786 kilometres above the earth's surface, 42,164.14 kilometres above the centre of the earth. This thin layer of spherical space where geosynchronous orbit can occur therefore has a surface area that can be calculated: it is slightly under 22.34 billion square kilometres.

i. Distributing 100,000 objects evenly across 22.34 billion square kilometres, how much space will each object theoretically have to itself?

a) 188,300 square kilometres
b) 223,400 square kilometres
c) 392,800 square kilometres

- *Area of Utah: 219,931 square kilometres*
- *Area of mainland Britain: 229,988 square kilometres*

A defunct satellite – one of the larger objects future space travellers might encounter – could be about the size of a small car. The risk of an accident involving space junk can therefore be expressed thus:

Imagine you were driving the length and breadth of Britain, with literally not a single other vehicle on the roads ... save that somewhere, perhaps on some motorway, perhaps down a country lane, perhaps in the northern reaches of Scotland, perhaps in the toe of Cornwall, there lurks a wrecked Mini Cooper, perhaps on the carriageway, perhaps off in a field beside the road, perhaps even off on some windswept hill, miles from the nearest tarmac. What do you think are the chances you might collide with it? In your car, which has, by the way, object detection radar filled as standard?

- *Total length of UK roads (1999): 371,913 kilometres*

Texas Tea

According to the CIA at the beginning of 2002 proven global oil reserves amounted to 1.025 trillion barrels.

Oil consumption worldwide is 75.81 million barrels a day.

Answer: i) b

Launching the space shuttle costs about $15,000 per payload kilo: less than gold or cocaine.

> ### i. How many days will the world's oil last at this level of consumption?
>
> a) 4,873.992
> b) 13,520.643
> c) 9,225.337

> ### ii. Starting from midnight on 31 December 2001, if oil consumption continues at the current rate at what time on what date will the world's oil supply run out?
>
> a) 3:25:55 p.m. on 7 January 2039
> b) 12:45:33 p.m. on 19 August 2023
> c) 5:17:02 p.m. on 23 October 2055

Number Freaking cannot predict what the last drop will be used for. The world's last lipstick perhaps?

Lippy

An average British woman uses 6 pounds (2.72 kilograms) of lipstick during a lifetime – a lifetime she can now expect to last 80 years. Make the Number Freaking assumption that she first puts on lipstick on the morning of her fifteenth birthday, then wears it every day until she dies on the morning of her eightieth.

Answers: i) b ii) a

The cab fare from New York City to Los Angeles including a 10 per cent tip would be $9,160.

---

i. How much lipstick does an average British woman use in a day?

a) 1.33 grams
b) 0.11 grams
c) 0.64 grams

There are 26.32 million British women aged fifteen plus.

---

ii. How much lipstick do British women use between them in a day?

a) 1.43 tonnes
b) 2.90 tonnes
c) 7.30 tonnes

And how many lipsticks must die to make this so? Number Freaking does not depend on extensive research. A department store lipstick weighs 3.7 grams. British women therefore discard 784,000 lipstick cases every day.

You Say Chips, I Say Crisps

Apparently Britons are the world's greatest crisp fiends, consuming a total of 12 billion packets a year.

Assume the British population to be 60 million.

---

i. How many packets of crisps is that per head per year?

a) 170
b) 185
c) 200

Answers: i) b ii) b i) c

Eight hundred Americans a week are injured by their jewellery.

Crisps come in various sizes. You may have noticed how smaller sizes are slowly being replaced by bigger sizes. At 50 grams per packet, individual consumption rises to around 10 kilograms per head – or 600,000 tonnes nationally.

- *Total annual potato production in Ireland: 600,000 tonnes*

You're Supposed to Eat It With a Spoon

Give a honeybee a flower and it'll collect pollen. Give it a tree and it will collect propolis, a gum which pundits at the chat show end of medicine (in Taiwan) have recommended for SARS – creating a demand which beekeepers in South East Asia have simply been unable to meet.

More traditionally, a colony of honeybees will produce about 60 pounds (27.2 kilograms) of honey a year. Throughout the season a colony might contain up to 60,000 animals.

i. How many grams of honey is that per bee?

a) 1.478
b) 0.954
c) 0.453

ii. How many bees does it take to produce a 227-gram jar?

a) 501
b) 878
c) 1,383

Answers: i) c ii) a

It takes 23 gallons of water to produce a pound of tomatoes.

Global honey production is about 1.2 million metric tonnes annually.

iii. So how big is the global honeybee swarm?

a) 724 billion
b) 4.33 trillion (1,330,000,000,000)
c) 2.65 trillion (2,650,000,000,000)

- *Total global cattle herd: 1,294,604,000*
- *Total global sheep herd: 1,202,920,000*
- *Total global pig herd: 857,099,000*
- *Total global goat herd: 94,266,000*
- *Total global camel herd: 19,627,000*

Let's Have a Whip Round

On average people in the UK each have £8.45 ($15.18) in small change.

If we just consider the UK's 35.5 million adults ...

i. How much do British adults possess in total in small change?

a) £310,538,237 ($557,913,000)
b) £299,975,000 ($538,935,080)
c) £276,653,737 ($497,036,090)

- *Estimated personal wealth of easyJet owner Stelios Haji-Ioannou: £300 million*

The average American carries $104 in their pocket.

Answers: iii) c i) b

Bees produce enough wax for every 10 grams of honey to make a birthday cake candle.

There are about 200 million Americans aged between 15 and 64.

---

ii. How much cash do 15–64-year-old Americans carry around between them?

a) $20,800,000,000 (£11,577,424,000)
b) $18,600,000,000 (£10,352,888,000)
c) $24,100,000,000 (£13,414,226,000)

- *Estimated personal wealth of Microsoft co-founder Paul Allen: $20.48 billion (£11.4 billion)*

Gorilla Marketing

Like a misquote from that song about a teddy bears' picnic, visitors casually smoking at Zhengzhou Zoo in China's Henan province are in for a big surprise: they're being panhandled for their unfinished butts by a thirteen-year-old captive female chimp named Feili, who China's Xinhua news agency reports has just taken up smoking.

The total number of chimps left on earth is estimated to be no more than 150,000.

- *Population of Aurora, Illinois: 150,000*

So imagine that every chimpanzee on earth started to smoke the Chinese smokers' average of 14.5 cigarettes a day – at a premium Western price of £4.50 for a packet of 20. (We shall not worry where they might get the money from, or indeed how they might get to the shops to buy them.)

Answer: ii) a

The Belgian 1 euro coin is reported to land 'heads up' 56 per cent of the time when it is tossed.

> **i. What would be the value of the world chimp market for cigarettes?**
>
> a) £178.74 million ($321.1 million) per year
> b) £234.88 million ($422 million) per year
> c) £90.22 million ($162.1 million) per year

Chimps would probably be less worried than humans about the health risks. Among humans a 45-year-old smoker has twice the chance of being dead by 60 than a non-smoker. Among chimps their biggest threat is a poacher with an AK47 (the world's favourite assault rifle).

But why has Feili started? Her keeper thinks she's done it out of sexual frustration. Apparently her 41-year-old mate is having problems meeting her needs. Some female readers may know how she feels.

Creeping Crawlers

As any parent knows, kids are incredibly time consuming, often in ways no parent ever expects. For example, as they get older, some can find it extremely difficult to sleep alone, and demand a parent stay with them until they fall asleep – often for years ...

> **i. Imagine you spend 80 minutes a night staying with a child, watching them fall asleep, for seven and a half years. How many hours is this in total?**
>
> a) 3,652.5
> b) 2,466.67
> c) 4,812.75

Answer: i) a i) a

A healthy chimpanzee can live for 50 years.

ii. If a working week is 47 hours, how many working weeks is this?

a) 77.71
b) 43.76
c) 111.12

iii. Assume an average working year of 48 weeks, and an average annual salary of £25,170 ($45,220). How much might this 'duty' cost a parent who could have been 'moonlighting' instead?

a) £23,806.67 ($42,771)
b) £40,749.18 ($73,209)
c) £67,180.00 ($120,695)

And all this is in addition to the direct costs of parenting – now estimated at £80,000 ($144,000) per child.

Eat the Rich

At the time of writing, the average wage in the UK is £25,170 a year (£484 per week for 52 weeks), and the US dollar is trading against sterling at $1.7966 to £1.

i. Roughly how long does it take Bill Gates – earning $55.05 a second every second – to earn a British worker's average weekly salary?

a) 26 seconds
b) 16 seconds
c) 42 seconds

Answers: ii) a iii) b i) b

In the first year of life children grow about 25 centimetres on average.

For this money the average worker in the UK is currently working an average of 47 hours a week. And of course they pay tax. If we assume the taxman takes a total of 25 per cent, this reduces the time Bill must work to equal it to 12 seconds. (So Bill earns the average weekly take-home pay of five British people in full-time work every minute.)

---

ii. If the rate is £484 for 47 hours, how much does an average British worker earn in an hour?

a) £19.10
b) £14.45
c) £10.30

But of course for that wage the worker has to survive for every second of every day.

---

iii. So what is their real income per hour?

a) £1.09
b) £2.88
c) £4.63

---

iv. And per second?

a) 0.240 pence (0.43 US cents)
b) 0.160 pence (0.29 US cents)
c) 0.080 pence (0.14 US cents)

To put this number in perspective, if Bill Gates' fortune grows at $55.05 (£30.64) a second, he earns a multiple of the UK average wage of 38,300.

Answers: ii) c iii) b iv) c

v. The average price of a house in the UK is currently £177,474 ($318,850). So to have the same impact on the wealth of the average person in work in the UK as this price would have on Bill Gates' wealth, how much would the average house need to cost?

a) £4.63
b) £5.41
c) £3.71

Which at current prices in the UK is a little more than the £4.39 price of a meal at McDonalds.

Including the Pictures of Your Butt

In 2001 the UK consumed 12.6 million tonnes of paper and board. For expediency let's translate all of it into photocopier paper. Photocopier paper usually has a density of 80 grams per square metre.

i. How many square metres would 12.6 million tonnes of photocopier paper cover?

a) 163.5 billion square metres
b) 145.2 billion square metres
c) 157.5 billion square metres

ii. How many square kilometres would 12.6 million tonnes of photocopier paper cover?

a) 163,500
b) 145,200
c) 157,500

Answers: v) a i) c ii) c

US dollar bills are made of 25 per cent linen and 75 per cent cotton.

- *Combined area of England and Wales: 151,243 square kilometres*

The UK has a population of 59.66 million people.

> **iii. Expressing the board and paper they all use as photocopier paper, how many square metres of photocopier paper does each person in the UK use on average each year?**
>
> a) 3,256.2 square metres (57.06 metres square)
> b) 2,640 square metres (51.38 metres square)
> c) 1,445.2 square metres (38.02 metres square)

- *Area of an Olympic-sized swimming pool: 1,000 square metres*

Who's Milking Who?

> **i. The average (Holstein) milk cow produces 7,850 litres of milk a year. How many cows would it take to supply the amount of milk equivalent to what it costs to run the royal family (67.81 million litres)?**
>
> a) 4,823
> b) 8,638
> c) 12,123

(The US President's $400,000 basic salary is equivalent to the milk production of about 53 cows.)

At the time of writing the total combined UK dairy herd was 2,207,000 cows. The average size of actual herds in the UK was 85 animals.

Answers: iii) b i) b

In 2001 just 4.9 million tonnes of paper and board was subsequently recycled in the UK.

ii. Assuming we all drink milk (just for a minute let's all be lactose tolerant), how many people share each cow?

a) 27
b) 73
c) 42

iii. In fact there are estimated to be 24.83 million households in the UK. So how many ordinary households share each cow?

a) 11.25
b) 5.9
c) 82.9

So, you share your cow with ten other households. The Queen's household gets 8,638 cows to itself. (Which is equivalent to nearly 102 average herds.)

• *A milking cow needs 32,809 square metres of grass to eat every year.*

It's a Privilege to Serve, Ma'am

The reported expense of the Queen's household is £36,392,600 per year. At current exchange rates this is equivalent to $65,382,945. A million dollars in single dollar bills weighs 1 metric tonne. It occupies a space of 1.129 cubic metres.

An industrial wheelbarrow has a piled capacity of 0.15 cubic metres and is 1.52 metres long.

Answers: ii) a iii) a

To produce a litre of milk a cow must consume 7.5 litres of water.

> **i. If it arrived in single US dollars how many wheelbarrows full of cash (by volume) would the Queen receive from the British taxpayer every year?**
>
> a) 492
> b) 823
> c) 1,453

Each of these barrows would be loaded with 150 kilograms of cash – over 300 pounds – far too heavy for one person to shift unaided under most Health and Safety regulations.

Spaced at 2-metre intervals this line of 492 wheelbarrows would stretch nearly a kilometre down the length of the Mall, the thoroughfare leading to Buckingham Palace.

Of course, if her income ever arrived this way, as a responsible employer the Queen would no doubt use liveried footmen working in teams to drag them into the palace.

Will Exchange Girlfriend for Beer

Having Number Freaked earlier around the hypothetical trade in boyfriends for chocolate, simple justice demands we now consider the trade in beer for girlfriends (as offered in the tee shirt slogan above).

So how much is a girlfriend worth? Broadly we can apply the same rules as with boyfriends and chocolate.

So, if she's a high-maintenance nag, a half of bitter at £1.50 ($2.69) is probably fair exchange. Alternatively, if she's A Lovely Girl Who Deserves Better, then as with a halfway decent boyfriend we'll base the deal on her body weight.

Answer: i) a

Winnie the Pooh earns Disney $5.9 billion annually.

The average woman in the UK now weighs 65 kilograms. Sixty-five kilograms of pils lager occupies a volume of 64.48 litres (113.46 UK pints; 14.18 UK gallons).

---

i. At £3 a pint how much is a 65-kilogram girlfriend worth?

a) £293.67 ($527.61)
b) £340.38 ($611.53)
c) £456.22 ($819.64)

A boyfriend, you may remember, was worth £396 ($711) under his 'swap for chocolate' deal, but his premium is entirely due to his extra weight. Weight for weight a girlfriend would be worth £414.74 ($745.12).

The official definition of binge drinking is five standard drinks, which as beer is 2.5 pints. The beer equivalent of a girlfriend is therefore enough to get an average man drunk 45 times. On average men get drunk 47 times a year. So exchanging A Lovely Girl Who Deserves Better for beer means you'll be drowning your sorrows for almost a year.

(To get a full year's beer in exchange, trade a girlfriend who's a couple of kilos overweight.)

Not All Girlfriends Are Equal Either

Just as some boyfriends are worth keeping, so too are some girlfriends. When considering a good one, again we must make some assumptions. The average wage for women in the UK is £20,314 ($36,496) per annum. Like men of the same age, a woman of 25 has a potential working life of 40 years. Unlike men of the same age, however, it is unlikely that her life will unfold with quite the same predictability.

Answer: i) b

There are 75 withdrawals from UK cash machines every second.

We'll work on the following assumptions. From 25 to 30 she'll earn the average wage. From 30 to 40 she'll be obligated by child care and earn nothing. Returning to part-time work, from 40 to 45 she'll earn one-third of the average wage (£6,771.33). At 45 she'll start her own business (the majority of starts-ups are now by women) and for ten years she'll earn the average wage again. Then for five years she'll do well and earn the white-collar average (£42,900). But just as things are looking up, from 60 onwards she'll have to stop work to nurse her partner, who's been pole-axed by a heart attack from overwork.

i. What, then, is such a woman's earning potential at current prices?

a) £624,093 ($1,121,245)
b) £732,002 ($1,315,115)
c) £553,067 ($993,640)

This is about 40 per cent of the earning potential of a comparable man – some £886,000 less. However, we shall not assume this to be the true lifetime value of a girlfriend. Why? Because, on average, women in the UK will, along the way, give birth to 1.9 kids each, which, as all parents know, are worth at least a million dollars, pounds or euros each …

Eek. Now we're Number Freaking in the land of the schmaltzy. Let's move on …

ii. At £3 a pint, how many pints of beer could you buy with £553,067?

a) 424,584
b) 184,356
c) 234,456

Answers: i) c ii) b

(Double this is the number of 'standard drinks' you could buy, of which five would get you legally drunk.)

> **iii. For how many years could this amount of beer supply you with five standard drinks a night?**
>
> a) 201.9
> b) 293.3
> c) 511.9

That's close to the 199 years it would take to eat a decent boyfriend's worth of chocolate ... But as any drunk who's ever taken a swig of last night's beer knows, beer quickly goes off: the shelf life for draught lager is normally about 60 days, for bottled lager about 110 days.

> **iv. How many people could get five-drink drunk daily for 60 days on this much beer?**
>
> a) 2,782
> b) 1,229
> c) 562

Now that would be a party ...

My Other Girlfriend's a Diamond

But what about a girlfriend versus other commodities? A 65-kilogram woman weighs 2,089.33 troy ounces. The price of gold at the time of writing is £226.82 per troy ounce in London and $404.30 in New York.

Answers: iii) a iv) b

The longest bar in America is 684 feet (208.5 metres) long.

> **i. How much would a 65-kilogram girlfriend's weight in gold cost (at London and New York prices)?**
>
> a) £473,902 ($844,716)
> b) £437,117 ($779,148)
> c) £352,758 ($628,781)

So given an earning potential of £553,067 ($993,640) – even without the child bearing – yes, a decent girlfriend is worth her weight in gold.

But what about in diamonds? Diamond prices vary by size, clarity, colour and cut. But as a guide, at the time of writing, a diamond trade web site is offering a one-carat 'brilliant cut' diamond for £8,957 ($16,092). One carat is a fifth of a gram. A one-carat diamond is 6.5 millimetres in diameter. (With £553,067 you could afford nearly 62.)

> **ii. How much is a 65-kilogram girlfriend worth, weight for weight, in one-carat 'brilliant cut' diamonds? ('Brilliant cut' is a technical term describing a round shape.)**
>
> a) £2.91 billion ($5.2 billion)
> b) £3.24 billion ($5.8 billion)
> c) £4.01 billion ($7.2 billion)

At this price Sir Richard Branson (net worth £2.6 billion) couldn't afford a girlfriend worth her weight in diamonds. Even the mighty Bill Gates could afford only half a dozen (5.96 to be exact). Bill is worth about $31 billion. So were his wife Melinda ever to divorce him, and get half his fortune, then assuming she weighed 65 kilograms at the time, she'd receive a settlement equivalent to about three times her body weight in one-carat diamonds.

But diamonds are made from rough diamonds. A two-carat rough sawable diamond (another technical description) can yield two half-carat cut

Answers: i) a ii) a

diamonds. To complicate matters two half-carat diamonds are worth nearly 60 per cent less than a single one-carat diamond. Half-carat (5.2 millimetre) brilliant cut diamonds can be had on the web for £1,912 ($3,435). So in half-carat ready cut diamonds a girlfriend would be worth a mere £1.24 billion ($2.2 billion). Sir Richard could afford two, Bill Gates, fourteen.

However, if you're still looking to save a few dollars, rough diamonds can be bought for as little as $200 (£111) per carat (in other words, for $400 per carat cut weight, in half-carat gems, if you organise your own cutting). £553,067 would buy 496 grams, over a pound in weight, of these.

> **iii. How much is a 65-kilogram girlfriend worth, weight for weight, in the rough diamonds you'd require to produce 65 kilograms of your own half-carat cut gems?**
>
> a) $110 million (£61.23 million)
> b) $120 million (£66.79 million)
> c) $130 million (£72.36 million)

Even Rod Stewart and Engelbert Humperdinck (net worth £75 million each) and Robbie Williams (net worth £78 million) could afford a rough diamond girlfriend. Which is presumably just what you'd expect. Just for the record, big Bill G. could afford 238, and Sir Richard B. could afford three dozen.

Always At It

Cynics suggest that newly married couples make love as often in the first year of marriage as they will for the whole of the rest of their lives.

How often a couple make love can depend on many things: from touchy-feely emotional stuff like trust, mood and intimacy, to no-nonsense practical stuff

Answer: iii) c
The Cullinan, the world's largest uncut rough diamond, weighs 3,106.75 carats.

like time, opportunity and fatigue. But there is one over-riding physical limitation – and that's down to the man. So for this reason we shall assume that during the first year of marriage a couple will make love three times a day.

The average age at which men in the UK marry nowadays is 30.5. Then they die at age 75. We shall therefore assume a married life span of 44.5 years.

> ### i. How many times will a couple make love during the first year of marriage?
>
> a) 887
> b) 1,096
> c) 1,329

> ### ii. So if the cynics are right, how many times per year will a couple make love for the remaining 43.5 years?
>
> a) 52.3
> b) 25.2
> c) 12.5

Which works out as equivalent to about once a fortnight.

QED.

Corner Café to the World

According to McDonald's' own figures, worldwide its staff serve 47 million customers daily in 31,000 restaurants. That's more people than are drunk worldwide at any given moment. More people than the population of Spain, more than the population of South Africa ...

Answers: i) b ii) b

On average it takes seven minutes to fall asleep.

i. The global population is 6.388 billion. At 47 million per day, how many days would it take for the world's population to get served?

a) 220 days
b) 182 days
c) 136 days

ii. If the entire world visited McDonald's, how many people would visit each store on average?

a) 435,005
b) 206,065
c) 787,339

iii. If all these 206,065 people visited a McDonald's store in 136 days, how many people would visit each day?

a) 4,361
b) 2,935
c) 1,515

iv. And if we assume local zoning regulations mean the average store is open twelve hours a day, how many customers must be served every minute?

a) 3.6
b) 8.2
c) 2.1

Answers: i) c ii) b iii) c iv) c

McDonald's employs over 1.5 million people in 119 countries worldwide.

A US advert for McDonald's famously claimed the chain would serve every customer within a minute. Thus to serve a meal to everyone on earth within this time frame, each restaurant would have to constantly employ three servers.

And every minute of every day 32,550 McDonald's meals would be being served.

● *Bill Gates could afford to buy a meal at McDonald's for everyone on earth.*

Roll Your Own

Sometimes Number Freaking requires virtually no data at all. We could go searching for appropriate data, or we could just wing it ...

So what did you have for breakfast?

Assume there are about 56 million people in Britain old enough or fit enough to eat solid foods. Assume one in four of them had cereal for breakfast. Why? Because, let us say, for argument's sake, there are four people in your household and one of them had cereal this morning. That's 14 million cereal breakfasts. Assume an average serving of cereal (regardless of variety) is about 30 grams.

> **i. Based on these figures, how many tonnes of cereal did Britons eat for breakfast this morning?**
>
> a) 870 tonnes
> b) 200 tonnes
> c) 420 tonnes

Answer: i) c

An average British man weighs £4,950 in fifty-pence pieces.

Cereal comes in boxes of various sizes. A popular size, you discover, consulting the side of a box, is 600 grams.

ii. Based on these figures, how many boxes of cereal did Britons eat for breakfast this morning?

a) 500,000
b) 600,000
c) 700,000

Of course, when Number Freaking, you may want to set your own parameters: you may not be interested in the UK – you may be interested only in, say, London. Well then, feel free to extemporise ... The population of London is about 7.3 million.

iii. So how many tonnes of cereal, based on these figures, did Londoners eat for breakfast this morning?

a) 54.75 tonnes
b) 72.75 tonnes
c) 90.50 tonnes

iv. And how many 600-gram boxes is that?

a) 45,050
b) 91,250
c) 189,300

You may calculate for yourself how many people found a plastic toy.

Answers: ii) c iii) a iv) b

Friends Reunited has over 11 million members. Allegedly, many want to share breakfast.

A Minute on the Tongue, a Lifetime on the Thighs

McDonald's serves 3 million customers daily in the UK. During a recent Parliamentary enquiry, it emerged that to work off the 'energy' in a McDonald's meal the average person would have to run 9 miles (14.4 kilometres).

i. The average person on earth has a life expectancy (at birth) of 67 years. Assuming that from the morning of their fifteenth birthday they ate only McDonald's meals for breakfast, lunch and dinner for the rest of their life – which Morgan Spurlock, for one, would not recommend – how many kilometres would they need to run, in total, to work it all off?

a) 820,498 kilometres (509,833 miles)
b) 1,057,180 kilometres (656,900 miles)
c) 2,763,788 kilometres (1,717,705 miles)

(Or slightly over an Olympic-distance marathon once a day every day – for life.)

The global equatorial circumference is 40,076 kilometres (24,903 miles).

ii. How many times would you need to girdle the earth to work off all these Big Mac meals?

a) 36.4
b) 20.6
c) 49.1

(Which, in an average lifetime, would mean going once around the world about once every two and a half years. Or put another way, the distance such a

Answers: i) a ii) b

It costs the price of over 200,000 McDonald's burgers to become a McDonald's UK franchisee.

McDonald's eater would have to run is the same as going to the moon and back 1.1 times.)

- *Big Mac meal: 1,411 calories*
- *Burger King Whopper and fries (US): 1,250 calories*
- *KFC meal: 910 calories*
- *Bacon sandwich: 730 calories*
- *Chicken curry: 650 calories*
- *Starbucks Grande Mocha (with whole milk and whipped cream): 400 calories*

The Student Diet

Dieticians, nutritionists and healthcare professionals all recommend diets that are healthy, balanced, low fat, high fibre, packed with fresh fruit and vegetables – and crunchy ... We, however, are Number Freaking: so we need not bother with issues such as saturated fats and free radicals, with Atkins or South Beach. Instead we shall cast ourselves back to a bygone era when all anyone worried about was raw calories.

Foodstuff	Portion	Calories
Bitter beer	1 pint	182
Guinness	1 pint	170
Lager (Stella Artois)	1 can (550 ml)	221
Wine	1 glass (120 ml)	87
Vodka	25 ml	55
Doughnut	49 g	140
Jaffa Cake	12 g	46
Scone	70 g	225
Bread, white	1 slice (37 g)	84
Toast	1 slice (33 g)	88

continued

Foodstuff	Portion	Calories
Cornflakes (Kelloggs)	45 g	167
Rice Krispies (Kelloggs)	45 g	171
Kebab	168 g	429
Pork sausage	1 sausage (24 g)	73
Bacon	1 rasher (25 g)	64
Beef in black bean sauce	386 g	432
Chicken and Cashew nuts	350 g	311
Egg fried rice	200 g	250
Cadbury's Creme Egg	39 g	174
Mars Bar	65 g	294
Chocolate	100 g	530
Kit Kat	2-finger bar (21 g)	106
Maltesers	37 g	183
Milky Way	26 g	117
Minstrels	42 g	209
Popcorn	100 g	405
Snickers	64.5 g	323
Twix	twin bar (62 g)	306
Cheese and onion crisps	35 g	184
Wotsits	40 g	212
Coffee	1 cup (220 ml)	15.4
Can of Coke	330 ml	139
Orange juice	1 glass (200 ml)	88
Tea	1 mug (270 ml)	29
Butter	10 g	74
Cheese, cheddar	40 g	172
Egg, size 3	57 g	84
Milk, whole	1 fl oz (30 ml)	20
Yoghurt, strawberry	1 pot (200 g)	123
Big Mac	215 g	492
Kentucky Fried Chicken	67 g	195
KFC fries	100 g	294
McDonald's fries	78 g	207

continued

Foodstuff	Portion	Calories
Banana	150 g	143
Apple	112 g	53
Chicken bhuna	300 g	396
Chicken korma	300 g	498
Chicken tikka	150 g	232
Poppadum	12 g	49
Samosa	50 g	126
Onion bhaji	22 g	65
Naan bread	half piece	269
Cheese and ham toastie	160 g	429
Chips	100 g	253

The maximum daily recommended calorie intake for men is 2,550, for women 1,940.

i. How many pints of Guinness need a man drink to reach his maximum daily calorie level?

a) 8
b) 15
c) 4

ii. On a toast mono-diet how many slices can a woman eat per day?

a) 12
b) 22
c) 7

Answers: i) b ii) a

iii. At college today you ate a Mars Bar, a strawberry yoghurt and a cheese and ham toastie. You drank two cups of coffee. Now you're off to the Union Bar. You've decided to have a kebab later for your tea. How many pints of bitter are you allowed?

a) 7
b) 2
c) 9

iv. You have a can of Coke, an apple and a packet of Wotsits in your bag. You put them on the table in front of you in the refectory, but when you come back from talking to your friend they've gone. All they have at the sales counter are Twix bars and bananas. How many Twix will you need to eat to replace what you lost?

a) 4.2
b) 2.4
c) 1.3

- *Recommended minimum daily intake of fruit and vegetables: 400 grams*

Fat Gits

It's estimated that half of the women in the UK are overweight or obese, and so are three-quarters of men. It's no wonder: we consume on average about 50 per cent more calories than we need.

Relative obesity is defined by a thing called body mass index (BMI). The 'healthy' range for BMI is between 18.9 and 24.9. A figure more or less means

Answers: iii) a iv) c

A man in the UK is 'allowed' 72.65 million calories in a lifetime, a woman 55.27 million.

you are over or under weight. A figure over 30 means you are obese. You can calculate your own BMI using this formula:

your weight in kilograms
_____ divided by _____
your height in metres multiplied by your height in metres

Or if you prefer:

your weight in pounds (multiplied by 703)
_____ divided by _____
your height in inches multiplied by your height in inches

In the UK the average man is 5 feet 9 inches (175.26 centimetres) tall and weighs 79.2 kilograms (almost 12.5 stone). He therefore has a BMI of 25.8.

The maximum healthy weight for a man of 5 feet 9 inches (i.e. the weight which gives a BMI of 24.9) is 76.483 kilograms (12 stone).

To bring his BMI back down to 24.9, an average 5 foot 9 inch man would need to grow to over 5 feet 10 inches (177.8 centimetres).

i. There are approximately 29.8 million adult men in the UK. On average, each of them is about 2.72 kilograms overweight. If all this extra weight could be converted into new men of 79.2 kilograms, how many extra men would this create?

a) 1,023,434
b) 2,250,889
c) 1,875,955

- *Population of Birmingham, UK: 977,000*
- *Size of the British Army at the outbreak of the First World War: 949,000*

Answer: i) a

In 1750 an average Englishman weighed 130 pounds (59 kilograms), and an average Frenchman weighed 110 pounds (50 kilograms).

Men in the UK are overweight by more than the population of Birmingham, the UK's second-largest city.

Yes Your Bum Does Look Big in That

As a measure of how tubby the British have become, in Yorkshire and Lincolnshire the average Body Mass Index (BMI) for men is now over 35, meaning men there of the UK average height of 5 feet 9 inches now weigh about 17 stone (108 kilograms). To be healthy and 17 stone you need to be 6 feet 10 inches tall.

i. If he had a BMI of 35 how tall would a man weighing 79.2 kilograms (the average weight for men in the UK) be?

a) 5 feet 6 inches
b) 4 feet 11 inches
c) 6 feet 1 inch

ii. If he had a BMI of 35 how tall would a man weighing 76.483 kilograms (the maximum healthy weight for a man of average height) be?

a) 5 feet 4 inches
b) 4 feet 10 inches
c) 5 feet 11 inches

Western medical textbooks often cite an average man as being 70 kilograms (11 stone).

Answers: i) b ii) b

In 1750, an average adult Englishman was 4 feet 7 inches tall.

iii. If he had a BMI of 35 how tall would a man of 70 kilograms be?

a) slightly over 5 feet 4 inches
b) slightly under 4 feet 8 inches
c) almost exactly 5 feet 11 inches

The message, then, for men in Yorkshire and Lincolnshire is GROW TALLER.

- *An average woman in the UK is 5 feet 4 inches tall, weighs 65 kilograms, and has a BMI of 24.6.*

A Sex Diet

An average man in Britain is about 2.71 kilograms overweight. You need to burn off 8,050 calories to lose 1 kilogram in weight.

On average a man burns off 336 calories during an hour making love.

i. How many extra hours' lovemaking should an average British man engage in to get back into shape?

a) 65
b) 122
c) 87

How often do women say they make love in a month?			
Age	1–2 times	3–12 times	13+ times
20–40	14%	55%	31%
41–60	28%	48%	24%
61–80	59%	29%	12%

Answers: iii) b i) a

Two hours making out = one McDonald's Big Mac.

Get a Room

Calories Burned Per Hour			
	Making the Beast with Two Backs	Making Out	Walking at 5 kph
Average woman	288	300	296
Average man	336	348	348
Average person	312	324	322

You and your latest squeeze are walking home from the pub. Normally it is a 15-minute walk. Tonight, however, in that passionate, heart-stopping, delirious way of a new relationship, you walk for three minutes, then stop off for five minutes of serious fooling around, then walk for three minutes, then get physical for another five, etcetera, all the way home ... where, upon arrival, you immediately tear each other's clothes off and rut like rabbits for ten minutes of torrid mutual delight.

i. How many calories would this disgusting (your mother said to say that) behaviour burn off (for an average person)?

a) 240.5
b) 466.9
c) 322.2

A couple of weeks later and you've both regained a little of your equilibrium. This evening, instead, you spend a delightful 45 minutes getting down and dirty on the sofa before retiring to the bedroom for an hour of languorous lovemaking.

Answer: i) a

> **ii. How many calories would this sort of nonsense consume (for an average person)?**
>
> a) 356
> b) 848
> c) 555

The King Is Dead

When he died Elvis weighed 159 kilograms and was eating 65,000 calories a day.

Put another way, he was about 80 kilograms overweight and eating in excess of 62,000 calories more than a man's recommended daily maximum.

You need to burn off 8,050 calories to lose one kilogram in weight. For a man of 159 kilograms, making love burns off 580 calories per hour.

> **i. How many hours' lovemaking would it have taken Mr Presley to work off a day's excess calories?**
>
> a) 107
> b) 143
> c) 176

Of course not even Elvis could have kept going for that long. In fact Americans reputedly average 28 minutes per coupling (call it half an hour), and few men could consistently maintain a sexual frequency of more than three times daily. Since this was the King, we'll assume he could manage it four times a day.

Answers: ii) c i) a
Unhooking a bra with two hands burns 8 calories.

ii. For how many days would Elvis have had to make love as often as he could (see above) to lose 80 kilograms?

a) 183
b) 365
c) 555

(That's about a week over 18 months.)

So maybe it wasn't a chronic lack of cycling that killed Elvis. Maybe it was a chronic lack of sex ...

Answer: ii) c

Unhooking a bra with one hand burns 18 calories.

If you enjoyed *Number Freaking*, you may also like …

50 Facts that Should Change the World

Jessica Williams

- Cars kill two people every minute

- More than 150 countries use torture

- America spends the same on pornography as it does on foreign aid

Think you know what's going on in the world?

Jessica Williams will make you think again.

Read about hunger, poverty, material and emotional deprivation, human rights abuses, unimaginable wealth, the decline of religion, the unstoppable rise of consumerism, mental illness, the drugs trade, corruption, gun culture, the abuse of our environment and much more in this shocking bestseller.

'Lucidly written, excellently researched, and with detailed referencing, the world won't look so rosy when you've put it down.' *Ecologist*

'A book to surprise, enrage and inform, it is a powerful antidote to apathy which offers information on how to make a difference. A gem of a book.' *Agenda*

'Provides proof of why we cannot be complacent about the world as it is today. Should become the bible of political activists everywhere.' *New Statesman*

A third of the world is at war
30 million people in Africa are HIV-positive
The US owes the UN $1 billion in unpaid dues
One in five people live on less than $1 a day

50

facts that should change the world
Jessica Williams

'A research handbook for the
No Logo generation'
Guardian

'Fearless and compelling. You need
to know what's in this book.'
Monica Ali

UK £6.99 * 1 84046 646 4

Introducing ...

Introducing is the highly acclaimed series of illustrated guides to big thinkers and subjects. Combining authoritative, accessible text by experts in their fields and witty, intelligent illustration by leading graphic artists, *Introducing* is a uniquely brilliant way to get your head round some challenging but fascinating ideas.

'*Introducing* is a miracle of modern publishing. Its gift has been to raise the précis to the level of an art form.' *Guardian*

'Often imitated ... seldom equalled' John Gribbin

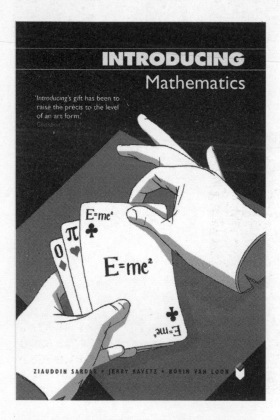

1 84046 637 5

1 84046 639 1

THE INTERNATIONAL BESTSELLER

INTRODUCING
Quantum Theory

'Introducing is a miracle of modern
publishing ... buy one now'

J.P. McEVOY • OSCAR ZARATE

1 84046 577 8

1 84046 581 6

Other titles:

Einstein	1 84046 667 7
Evolution	1 84046 634 0
Science	1 84046 358 9
Time	1 84046 592 1
The Universe	1 84046 068 7

All books £9.99

The Man Who Ate Bluebottles
… and Other Great British Eccentrics

Catherine Caufield
Illustrated by Peter Till

Until he ate a bluebottle, William Buckland had always maintained that the taste of mole was the most repulsive he knew. But that was before he ate the embalmed heart of Louis XVI.

Lord Monboddo believed that babies are born with tails, and was a careful observer at the births of his own children – but in each case the midwife outwitted him and managed to destroy the evidence.

These are just two of the eccentrics who enliven the pages of this delightful survey of British loopiness through the ages, a celebration of true originals whose strength of character stands out now more than ever in our age of mass-market conformity. As John Stuart Mill warned as long ago as the 1850s: 'That so few now dare to be eccentric marks the chief danger of the time.'

£9.99 * 1 84046 697 9

The Z–Z of Great Britain

Dixe Wills

From Zawn Organ,
Cornwall, to Zoar in
Shetland via Zion Place
in Somerset and Zulu Farm,
Oxfordshire, *The Z–Z of
Great Britain* is, arguably,
the definitive work on every
town, village and speck
on the map that begins
with a zed.

Providing an essential
profile of each and every
zed locale in the land and,
thoughtfully, instructions
on how to get there,
Dixe Wills has written the
new zedthusiast bible.

Z–Z also includes detailed maps and a unique rating system to
cut the hassle of planning your next zedventure.

And, for when you're enjoying a pint at the end of the day in
The Half Moon, Zig Zag Hill's local, *Z–Z* features a dazzling
introduction to the tempestuous and often distasteful history of
the letter at the very extremity of the English language …

UK £9.99 * 1 84046 689 8

Sex, Botany and Empire
The Story of Carl Linnaeus and Joseph Banks

Patricia Fara

When the imperial explorer
James Cook returned from his
first voyage to Australia, scandal
writers mercilessly satirised the
amorous exploits of his botanist,
Joseph Banks. Was the pursuit
of scientific truth really what drove
Enlightenment science?

UK £6.99 * 1 84046 573 5

Fatal Attraction
Magnetic Mysteries of the Enlightenment

Patricia Fara

Fatal Attraction tells the story
of magnetism in the Age of
Enlightenment. From
compasses to scientific
experiments, and from magic
tricks to Mesmerism, Patricia
Fara charts the 18th century's
fascination with nature's
strangest power.

UK £9.99 * 1 84046 632 4